UNDERSEA VEHICLES AND NATIONAL NEEDS

Committee on Undersea Vehicles and National Needs

Marine Board
Commission on Engineering and Technical Systems
National Research Council

NATIONAL ACADEMY PRESS
Washington, D.C. 1996

NATIONAL ACADEMY PRESS • 2101 Constitution Avenue, N.W. • Washington, DC 20418

NOTICE: The project that is the subject of this report was approved by the Governing Board of the National Research Council, whose members are drawn from the councils of the National Academy of Sciences, the National Academy of Engineering, and the Institute of Medicine. The members of the committee responsible for the report were chosen for their special competencies and with regard for appropriate balance.

This report has been reviewed by a group other than the authors according to procedures approved by a Report Review Committee consisting of members of the National Academy of Sciences, the National Academy of Engineering, and the Institute of Medicine.

The National Academy of Sciences is a private, nonprofit, self-perpetuating society of distinguished scholars engaged in scientific and engineering research, dedicated to the furtherance of science and technology and to their use for the general welfare. Upon the authority of the charter granted to it by the Congress in 1863, the Academy has a mandate that requires it to advise the federal government on scientific and technical matters. Dr. Bruce Alberts is president of the National Academy of Sciences.

The National Academy of Engineering was established in 1964, under the charter of the National Academy of Sciences, as a parallel organization of outstanding engineers. It is autonomous in its administration and in the selection of its members, sharing with the National Academy of Sciences the responsibility for advising the federal government. The National Academy of Engineering also sponsors engineering programs aimed at meeting national needs, encourages education and research, and recognizes the superior achievements of engineers. Dr. William A. Wulf is interim president of the National Academy of Engineering.

The Institute of Medicine was established in 1970 by the National Academy of Sciences to secure the services of eminent members of appropriate professions in the examination of policy matters pertaining to the health of the public. The Institute acts under the responsibility given to the National Academy of Sciences by its congressional charter to be an adviser to the federal government and, upon its own initiative, to identify issues of medical care, research, and education. Dr. Kenneth I. Shine is president of the Institute of Medicine.

The National Research Council was organized by the National Academy of Sciences in 1916 to associate the broad community of science and technology with the Academy's purposes of furthering knowledge and advising the federal government. Functioning in accordance with general policies determined by the Academy, the Council has become the principal operating agency of both the National Academy of Sciences and the National Academy of Engineering in providing services to the government, the public, and the scientific and engineering communities. The Council is administered jointly by both Academies and the Institute of Medicine. Dr. Bruce Alberts and Dr. William A. Wulf are chairman and interim vice chairman, respectively, of the National Research Council.

The program described in this report is supported by cooperative agreement No. 14-35-0001-30475 between the Minerals Management Service of the U.S. Department of the Interior and the National Academy of Sciences, by interagency cooperative agreement No. DTMA91-94-G-00003 between the Maritime Administration of the Department of Transportation and the National Academy of Sciences, and by grant No. N00014-95-1-1205 between the U.S. Department of the Navy, Office of Naval Research and the National Academy of Sciences.

Limited copies are available from:
Marine Board
Commission on Engineering and Technical Systems
National Research Council
2101 Constitution Avenue
Washington, D.C. 20418

Library of Congress Catalog Card Number 96-71679
International Standard Book Number 0-309-05384-6

Cover Photos:
Top: A picture of *Alvin,* a deep submergence vehicle. Photo based on a painting by George Warren Delano commissioned by Woods Hole Oceanographic Institute. Center: *Triton 19,* a remotely operated vehicle. Photo courtesy of Perry Tritech. Bottom: *ABE* (autonomous benthic explorer), an autonomous underwater vehicle. Photo courtesy of Woods Hole Oceanographic Institute.

Copyright 1996 by the National Academy of Sciences. All Rights Reserved.

Printed in the United States of America

Dedication

Mr. Howard R. Talkington was a member of this committee until his death in December 1993. Howard Talkington was a recognized world authority on ocean engineering and undersea vehicles. As a leading figure in the Navy Research Laboratory community, he was directly responsible for developing more than 30 different vehicle systems and much of the technology now used for operations to the depths of the abyssal plains. His quiet demeanor, his innovation, and his passion for ocean pioneering earned him the admiration of his colleagues everywhere, and he will be sorely missed. Howard's contributions to the early phases of this report were instrumental, and the report is respectfully dedicated to his memory.

COMMITTEE ON UNDERSEA VEHICLES AND NATIONAL NEEDS

J.B. "BRAD" MOONEY, JR., **NAE**, *chair*, U.S. Navy (retired), Arlington, Virginia
JOHN R. APEL, Johns Hopkins University (retired), Silver Spring, Maryland
ROBERT H. CANNON, JR., **NAE**, Stanford University, Stanford, California
JOHN R. DELANEY, University of Washington, Seattle
NORMAN B. ESTABROOK, SAIC Marine Engineering Group, Del Mar, California
LARRY L. GENTRY, Lockheed Martin (retired), Sunnyvale, California
JAMES R. MCFARLANE, International Submarine Engineering, Port Coquitlan, British Columbia, Canada
ANDREW L. "DREW" MICHEL, ROV Technologies, New Orleans, Louisiana
BRUCE H. ROBISON, Monterey Bay Aquarium Research Institute, Moss Landing, California
MARY I. SCRANTON, State University of New York, Stony Brook
PETER H. WIEBE, Woods Hole Oceanographic Institution, Woods Hole, Massachusetts
DANA R. YOERGER, Woods Hole Oceanographic Institution, Woods Hole, Massachusetts

Liaison Representatives

NORMAN CAPLAN, National Science Foundation
LARRY CLARK, National Science Foundation
RICHARD M. HAYES, Office of the Oceanographer of the Navy
AL KALVAITIS, National Oceanic and Atmospheric Administration
JAMES ANDREWS, Office of the Chief of Naval Research
EDWIN L. LANCASTER, U.S. Navy Deep Submergence Office
DONALD E. PRYOR, National Oceanic and Atmospheric Administration
CHARLES E. STUART, Advanced Research Projects Agency

Staff

DONALD W. PERKINS, Project officer
DELPHINE D. GLAZE, Administrative Assistant
WYETHA B. TURNEY, Production Assistant

MARINE BOARD

JAMES M. COLEMAN, **NAE**, *chair*, Louisiana State University, Baton Rouge
JERRY A. ASPLAND, *vice-chair*, California Maritime Academy, Vallejo, California
BERNHARD J. ABRAHAMSSON, University of Wisconsin, Superior
BROCK B. BERNSTEIN, EcoAnalysis, Ojai, California
LILLIAN C. BORRONE, **NAE**, Port Authority of New York and New Jersey, New York
SARAH CHASIS, Natural Resources Defense Council, New York
CHRYSSOSTOMOS CHRYSSOSTOMIDIS, Massachusetts Institute of Technology, Cambridge
BILIANA CICIN-SAIN, University of Delaware, Newark
BILLY L. EDGE, Texas A&M University, College Station
JOHN W. FARRINGTON, Woods Hole Oceanographic Institution, Woods Hole, Massachusetts
MARTHA GRABOWSKI, LeMoyne College and Rensselaer Polytechnic Institute, Cazenovia, New York
JAMES D. MURFF, Exxon Production Research Company, Houston, Texas
M. ELISABETH PATÉ-CORNELL, **NAE**, Stanford University, Stanford, California
DONALD W. PRITCHARD, **NAE**, State University of New York at Stony Brook, Severna Park, Maryland
STEVEN T. SCALZO, Foss Maritime Company, Seattle, Washington
MALCOLM L. SPAULDING, University of Rhode Island, Narragansett
KARL K. TUREKIAN, **NAS**, Yale University, New Haven, Connecticut
ROD VULOVIC, Sea-Land Service, Charlotte, North Carolina
E.G. "SKIP" WARD, Shell Offshore, Houston, Texas

Staff

CHARLES A. BOOKMAN, Director
DONALD W. PERKINS, Associate Director
DORIS C. HOLMES, Staff Associate

Preface

Undersea vehicles in a variety of forms have expanded human access to the deep, both for scientific research and for work such as cable-laying and offshore oil and gas operations. They will help answer new questions about the future of the human environment and the structure of the Earth. Information gathered by undersea vehicles has informed the public since the 1960s. News and film footage from undersea vehicles commanded by Jacques Cousteau, more recently the wreckage of the *R.M.S. Titanic*, the *Lusitania*, and the battleship *Bismarck*, and Roman ships on the bed of the Mediterranean off Sicily have received worldwide attention, thanks to the firsthand view offered by undersea vehicles. The activities of scientists involved in experiments near the Galapagos Islands off Peru were brought into the American classrooms on a real-time basis interactively bonding students and investigators in actual scientific investigations. Advances in the technology that made these activities possible suggest that future applications of underwater sensors, sonar, manipulators, probes, and samplers will be no less exciting.

Developments and improvements in underwater vehicle capabilities historically have been responsive to major national undersea sensing, monitoring, inspecting, and work-related requirements in both military and civil sectors. Remotely operated vehicles (ROVs) have largely replaced divers as offshore oil and gas operations (exploration and production) extend out into the Gulf of Mexico. They are also widely used in laying undersea cables. Industry will soon be capable of operating in depths down to 1,650 meters (5,400 feet)[1] of the Gulf of Mexico. Further operational extensions in distance from shore and in depth will place increasing demands on better ways to safely and reliably perform both routine and emergency work on or near the seafloor.

National interest in global environmental change, pollution monitoring and control, and use of undersea living and mineral resources has given impetus to developing a more comprehensive understanding of the oceans, requiring frequent in situ observations to characterize geological, biological, physical, and chemical phenomena. Experts have estimated that the kinds and quantities of data needed cannot be obtained in the necessary time frame using traditional methods. Similarly, present practices for exploring the seabed for minerals or performing bathymetry are too slow and expensive when performed using present ship and towed instrumentation; new types of platforms and tools are needed. Moreover, new enabling technologies will be needed to explore and describe the seas and seabed in remote areas, such as under the arctic ice cap and in the icebound regions of the oceans surrounding the Antarctic continent.

ORIGIN OF THE STUDY

The Marine Board of the National Research Council (NRC) has become increasingly aware of the national need in industry, government, and science to improve systems for doing undersea work and research. The Persian Gulf War reemphasized the need to develop underwater vehicles for coastal anti-mine warfare. The end of the Cold War resulted in an increased emphasis on dual military-civilian use of technology. This change has raised hopes for transfer of military developments and field experience to civilian applications. Accordingly, the Marine Board convened two planning meetings in 1991 to learn about recent developments and national needs in undersea technologies. A clear consensus emerged from these meetings that there is a need for a strategy to guide and encourage the development and deployment of undersea vehicles and their applications in response to national interests.

Although there had been some previous review efforts, each had its own constituency or mission to govern the outcome, and there has been no convergence of conclusions and recommendations over a broad set of findings. In response to the findings of the planning meetings, which called for a single assessment encompassing all types of undersea vehicles and all nonmilitary uses, five agencies—the Office of

---

[1] Vehicle systems are in design and under construction to support drilling and installation for the Shell Mensa project in the Gulf of Mexico, which will be operating of from 1,650 meters depth.

Naval Research and the Deep Submergence Office of the U.S. Navy, the National Science Foundation, the National Undersea Research Program of the National Oceanic and Atmospheric Administration, and the Advanced Research Programs Agency of the U.S. Department of Defense—provided funds to undertake this study. Accordingly, the NRC's Commission on Engineering and Technical Systems assembled a committee under the auspices of the Marine Board to design and recommend a national strategy for the development of undersea vehicles and their nonmilitary applications.

Committee members were selected for their expertise and to achieve balanced viewpoints (biographical information is presented in Appendix A). The principle guiding the constitution of the committee and its work, consistent with the NRC policy, was not to exclude any bias that might accompany expertise vital to the study but to seek balance and fair treatment. The resulting committee membership balanced the technical, scientific, and research management disciplines and recognized the industrial, university, and government concerns in its composition.

SCOPE OF THE STUDY

The task of the Committee on Undersea Vehicles and National Needs was to assess the needs of the nation in regard to the industrial and scientific requirements for acquisition of in situ oceanographic data and performance of undersea work tasks. The objective was to devise a development strategy with objectives, priorities, and guidance on implementation leading to improvements in undersea vehicle technology in all appropriate civil applications. All types of undersea vehicles, whether human-occupied (the term used by the committee in lieu of "manned") or unoccupied, and all civilian applications of vehicle platforms were to be considered in this assessment. The committee defined an undersea vehicle as: "A mobile, controlled, self-propelled subsurface platform capable of carrying sensors and tools." Three broad types of vehicles were distinguished in the study: human-occupied vehicles, commonly designated as deep submersible vehicles (DSVs); remotely operated vehicles (ROVs); and autonomous underwater vehicles (AUVs). This categorization allowed for consideration of hybrid vehicle systems such as a combination of an AUV with an acoustic or fiber-optic link to a surface ship or central control point. Specific tasks to be undertaken by the committee were as follows:

- Review and assess the status of vehicle system technology and development in the United States and foreign countries and assess military developments that have potential civilian applications.
- Identify and assess the needs for in situ oceanographic data and undersea work tasks that stem from specific engineering and scientific, national and international programs and goals; assess the role that undersea vehicle systems play in achieving these goals; and establish the required capabilities for vehicle systems. In the role assessment, identify unique functions and observations that can be enabled by the use of undersea vehicles that cannot be achieved by any other means and appraise the value of these unique capabilities for accomplishing national undersea missions.
- Compare the present state of vehicle technology, operational practice, and present research in relation to the national needs and the vehicle system capabilities that will be required to be responsive to those scientific and engineering needs.
- Identify costs of developing and operating vehicles and contrast these costs with other means of performing the same tasks.
- Recommend strategies for developing and applying undersea vehicle systems to enhance the efficient and economic means of acquiring data and performing undersea tasks, including recommendations for research and development.
- Provide guidance regarding the role of industry, academia, and government in fostering advances in vehicle system technology and its application.

In establishing the types of vehicle platforms that it would assess, the committee determined that military submarines would be excluded, even if they were to be used for civilian missions. The bases for this judgment were that these platforms are designed for military missions, and civilian scientific operations are short-term secondary missions; they are owned and operated by the U.S. Department of Defense; and they are nuclear-powered, thus requiring extensive and costly support systems. In short, military submarines are not vehicles used in industry, university, or other long-term, non-defense applications. The Navy's *NR-1*, although not designed as part of a weapons system, was also excluded from the scope of the assessment, except in cases where it might be useful in developing vehicle technology.

Tourist submersibles, while a substantial industry, were also considered to be outside the scope of this study because their objective includes neither work tasks nor sensing and observation for science. This is a large commercial activity involving more than 1 million passengers per year. A 1990 NRC report (*Safety of Tourist Submersibles*) provides extensive, useful information on the technology of this sector of the submarine and submersibles industries.

The committee reviewed capital and operating cost information about various types of vehicles and several alternative platforms for obtaining oceanographic data and working beneath the sea and determined that there was no satisfactory way to make cost comparisons, nor was sufficient cost information available.

STUDY METHODS AND REPORT ORGANIZATION

The committee met seven times during a three-year period beginning in October 1992. The committee received briefings from representatives of each of the sponsoring agencies in regard to their nonclassified development and interests in the application of undersea vehicles. Briefings concerning the state of practice and possible uses of vehicles in industry were given by experts in offshore oil and gas operations, undersea search and survey processes, and undersea telecommunications systems. Science objectives, involving data acquisition and work in the sea and on the seabed, were provided by leaders in chemical oceanography, marine biology, marine geology, atmospheric science and climatology, and physical oceanography. Vehicle development activities in universities and industry were identified and reviewed through presentations to the committee by research investigators, participation by committee members and staff at symposia held under the auspices of the Massachusetts Institute of Technology, and the 1993 and 1995 conferences on underwater intervention sponsored by the Marine Technology Society and the Association of Diving Contractors.

Following its first meeting in the information acquisition phase of the study, the committee organized itself into three panels: science applications, industry applications, and technology. These panels, which included experts in vehicle design and construction and in the oceanographic sciences, met individually on four occasions following committee meetings and in separate sessions.

Chapter 1 of this report surveys challenges before the United States in understanding the processes within the sea and the sea-atmosphere system and in developing and managing ocean and seabed resources. The role of undersea vehicles as a tool for industry and science in meeting that challenge is also introduced in Chapter 1. Chapter 2 describes the technology of each type of undersea vehicle and reviews and assesses the state of development and practice for the principal vehicle system subsystems. Chapter 3 presents science and resource development objectives and needs in terms that relate to opportunities for application of various vehicle systems. Four "focal projects" illustrate the potential link between available vehicle technology and possible areas for development that would enable undersea vehicles to be responsive to the emerging national needs within the sea and on the seabed. Chapter 4 synthesizes the possibilities for technical improvements and the potential for advancing vehicle mission capabilities that will be most useful in enhancing and enabling scientific missions and industrial tasks undersea. Chapter 5 addresses the trends affecting the nation's ability to work and measure in the sea, the choices for implementing change, and the benefits or consequences that might be expected from action or the lack of action. The committee's recommendations for development, implementation of investment, and improvement of access to vehicle resources are summarized in the Executive Summary and detailed in Chapter 6.

ACKNOWLEDGMENTS

The committee wishes to express its thanks to the many individuals who contributed their time and energy to this project, whether by assisting the committee on its panels or through presentations, correspondence, and telephoned inputs. Representatives of the sponsoring federal agencies and industry, oceanographers, undersea vehicle operators, and vehicle designers all provided invaluable assistance.

Particular appreciation goes to the liaison representatives of the project sponsors: Norman Caplan and Larry Clark, National Science Foundation; Marsh Youngbluth (until December 1993) and Al Kalvaitis of National Oceanic and Atmospheric Administration's National Undersea Research Program; Keith Kaulum (until August 1994), Office of Naval Research; Lieutenant Commander George Billy (until August 1994) and Lieutenant Commander Edwin Lancaster, U.S. Navy Deep Submergence Office (N879); and Captain Alan Beam, Advanced Research Projects Agency of the U.S. Department of Defense (until December 1994). Other agencies and office representatives also participated in several meetings, including Donald Pryor of the National Ocean Service of the National Oceanic and Atmospheric Administration and Richard Hayes, Office of the Oceanographer of the Navy.

Special thanks are due to Don Walsh, International Maritime Incorporated, who served as Marine Board liaison to the committee until July 1994. As a consultant, he provided significant information from his long and extensive experience as a vehicle pilot and a manager of diving operations. Claude Brancart, Draper Laboratory, also provided assistance in the work of the committee, particularly in support of AUV technology assessment. Others who assisted the technology panel of the committee were Craig Mullen, Oceaneering Technologies, Inc.; and John Jacobson, Perry Tritech; both of whom provided insight into the development and application of ROVs. D. Richard Blidberg, Marine Systems Engineering Laboratory of Northeastern University, provided significant background information concerning the development of AUVs and their applications, stemming from his long experience in vehicle engineering education projects, research, development, and field testing during his directorship of the Unmanned Underwater Vehicle Laboratory at the University of New Hampshire. Lynne Carter, executive director, Center for Ocean Management Studies, University of Rhode Island, served as a consultant during the early formulation of the report and worked very effectively with the science panel of the committee in developing Chapter 3 of the report. William Schroeder, the University of Alabama, and William Sprigg, director, Board on Atmospheric Sciences and Climate, NRC, assisted the committee through their contributions to the deliberations of the committee's science panel.

The chairman wishes to recognize the members of the committee for their commitment of time and for their individual efforts in gathering information for the study, reviewing and assessing information, and preparing drafts.

Contents

EXECUTIVE SUMMARY 1

1   INTRODUCTION 7
    Undersea Vehicles Defined, 8
    Evolution of Undersea Vehicles since the 1950s, 12
    U.S. Trends in Undersea Vehicle Development and Use, 15
    Foreign Programs, 16
    Findings, 16
    References, 17

2   UNDERSEA VEHICLE CAPABILITIES AND TECHNOLOGIES 18
    Vehicle Systems, 18
    Deep Submersible Vehicles, 19
    Remotely Operated Vehicles, 20
    Autonomous Underwater Vehicles, 21
    Operational Attributes of Vehicle Systems, 24
    Evaluation of the State of Technology, 25
    Systems Integration, 40
    Technology Transfer from Other Industries and Technical Fields, 41
    Findings, 42
    References, 43

3   VITAL NATIONAL NEEDS 47
    Scientific Understanding and Applications, 47
    Living Resources and Environmental Management, 55
    Marine Industrial Activities, 59
    National Security, Public Safety, and Regulation, 66
    Findings, 67
    References, 68

4   PRIORITIES FOR FUTURE DEVELOPMENT 70
    Important Needs and Opportunities, 70
    Setting Priorities for Undersea Vehicle Development, 71
    Findings, 74

5 ENHANCING THE NATION'S CAPACITY FOR UNDERSEA
   WORK AND RESEARCH 76
      Need for a Strategic Approach, 76
      Trends and Policy Alternatives, 78
      Findings, 81
      References, 81

6 CONCLUSIONS AND RECOMMENDATIONS 82

APPENDICES

A Biographical Sketches of Committee Members 87
B Foreign Developments 90
C Development of Deep Submersible Vehicles in the United States:
      1958–1994 93
D U.S. Government Agencies That Own and/or Use Submersibles 95
E Deep Submersible Vehicles in Service or Available Worldwide 96

ACRONYMS 99

List of Boxes, Figures, and Tables

BOXES

2-1 AUV Example: *AUSS*, 22
2-2 AUV Example: *Odyssey*, 23
2-3 AUV Example: *ABE*, 24
3-1 Focal Project 1: Synoptic Observation System, 52
3-2 Focal Project 2: Blue Water Oceanographic Sections and Hydrographic Surveys, 56
3-3 Focal Project 3: Subsea Oil Field Inspection and Intervention, 62
3-4 Focal Project 4: Search and Survey, 64

FIGURES

1-1 Photo of *Alvin*, 9
1-2 Photo of *Deep Rover*, 9
1-3 Photo of *Phantom*, 10
1-4 Photo of *Triton 19*, 10
1-5 Photo of *Ventana*, 10
1-6 Photo of *UUV II*, 11
1-7 Photo of autonomous benthic explorer *(ABE)*, 11
1-8 Photo of *Theseus*, 11
1-9 Photo of *Odyssey I*, 12
1-10 Photo of *Johnson Sea-Link*, 13
1-11 Photo of *Shinkai*, 14
1-12 Photo of *Kaiko*, 15
2-1 Schematic diagram of vehicle systems, 19
2-2 Battery cell comparisons, 28

TABLES

2-1 Comparative Undersea Vehicle Capabilities, 20
2-2 Current Undersea Vehicle Capabilities, 25
2-3 Performance Characteristics of Available Energy Sources, 27
2-4 Technology Transfer, 41
3-1 Focal Projects: Responses to National Needs, 48
4-1 Technology Assessment Summary by Subsystem, 71

Executive Summary

The Earth's oceans are the central engine of the energy and chemical balance that sustains humankind. They provide warmth and power. They moderate the weather so food can be grown on land to feed the Earth's population. Their living resources also supply food. Understanding them is one of humanity's most important challenges.

The United States, as a leading maritime nation, relies on the ocean more than most. It has one-fourth of the world's trade and the largest standing navy. Investments in ocean science and technology by U.S. government and industry lead the world. But support for those investments has waned steadily in recent decades, even as the need for knowledge of the ocean, in all of its dimensions, has grown. The nation faces decisions that will require substantially more information about pollutant concentrations, climate change, the ocean food chain, and the hydrocarbon and mineral resources beneath the ocean and on the bottom.

Humanity's first scientifically based knowledge of the ocean came from observations made aboard surface vessels and from instruments put into the sea. Later came aircraft and satellites with sensors aboard. Underwater vehicles augment these capabilities by making it possible to go far beneath the surface, giving human beings firsthand information about how the oceans work. Early bathyspheres took explorers to spectacular depths. Four decades of exploration with undersea craft have since offered glimpses of many facets of the oceans' wonders, suggesting how much there is yet to discover. Recent oceanographic and geologic discoveries of immense significance made while using undersea vehicles, such as the stunning discovery of hydrothermal vents at midocean plate boundaries with associated chemosynthetic food chains, are tantalizing examples of how much more there is to learn about the sea.

Advances in guidance and control, communications, sensors, and other technologies for undersea vehicles offer an opportunity for a great leap in understanding the ocean engine. Together these advances have led to the evolution of remotely piloted and autonomous undersea vehicles, which may extend the human reach beneath the sea, offering more cost-effective and capable systems for some missions than those used today.

This report, developed by a balanced team of ocean scientists, underwater vehicle development engineers, and managers in both government and industry, assesses the value of investments in underwater vehicles. On the basis of that assessment, the committee offers recommendations for definitive actions that will lead to systems that offer access to knowledge and tools that protect and enhance human life.

Undersea vehicles are of three types:

- Deep submersible vehicles (DSVs) are human occupied and are generally used to descend to great depths (up to 6,500 meters), with up to several kilometers of horizontal mobility on the bottom. They are used for a variety of tasks requiring human observation or recovery of objects or samples. They have relative brief endurance (owing to human limitations) and are relatively costly to operate and maintain, in part due to provisions to ensure human safety.
- Remotely operated vehicles (ROVs) are unoccupied, tethered vehicles with umbilical cables to carry power, sensor data, and control commands from operators on the surface. ROVs are widely used in the offshore oil and gas industry for a variety of inspection and manipulation tasks and in laying undersea cable. They are also used in ocean research. ROVs are maneuverable within the limits of their tethers (a radius of up to a kilometer) and have nearly unlimited endurance on the bottom. Large ROVs used for research commonly operate from dedicated surface support ships. (In industry, they may be operated from construction or production vessels or platforms using temporary launching and retrieval equipment.)
- Autonomous underwater vehicles (AUVs) are unoccupied submersibles, without tethers, powered by onboard batteries, fuel cells, or other energy sources. Currently, they are intended to carry out preprogrammed missions, with little or no direct human intervention. The ability

to achieve human strategic control of AUVs in near-real-time is now within reach. Still experimental, AUVs have yet to see wide operational use.

The capabilities of these vehicles have grown dramatically in the past two decades or so. In particular, advances in sensors, control systems, communications, and manipulators have made ROVs increasingly strong and versatile performers to the extent that they have largely replaced DSVs in commercial operations. These technical advances also have opened the potential for operational AUVs able to carry out complex and precise survey or sampling missions. DSV technology is considered generally mature. AUVs offer the greatest potential for expansion of capabilities as the result of investments in development, although both ROVs and DSVs will remain important tools of science and industry. AUVs promise to open new areas of research and new ways of doing work undersea, if their development is supported. Many of the technology developments carried out for AUVs will be applicable to other vehicle types.

The United States was globally preeminent in undersea vehicle technology in the 1960s and 1970s, but today there is no concerted government program to develop and use these systems. Other nations, including Japan, Canada, France, the United Kingdom, and Russia, have carried the technologies forward. Since the early 1970s the United States Navy has not built a single DSV, and commercial interest in advancing the technology has faded. The ROV technology base has shifted from military programs to private industry, which develops the technology of these robust and versatile systems to meet the growing demands of the offshore oil and gas industry and other customers. There are now over a hundred work-class ROVs operating worldwide, and this count is continuing to increase at a rate of 10 or more per year. Developing major AUV technology remains the concern of government programs.

The nation still has a strong operational undersea vehicle capability, mainly in U.S. Navy hands. The vehicles are increasingly outdated but quite capable and varied. With the end of the Cold War, the Navy opened access to its DSVs more widely to scientific researchers. The Deep Submergence Science Committee—supported jointly by the National Undersea Research Program of the National Oceanic and Atmospheric Administration, the National Science Foundation, and the U.S. Navy—oversees the *Alvin* and *Medea-Jason* vehicle facility at the Woods Hole Oceanographic Institution and, as an added task, reviews proposals for the use of these vehicles as well as other government-owned vehicle assets used in nonmilitary research. The Deep Submergence Science Committee also assists in scheduling scientific missions using these assets. The U.S. Navy and National Science Foundation are responsible for the costs of running most of the vehicles. The Navy DSVs, *Turtle* and *Sea Cliff*, are now available for civilian purposes up to 60 days each year. The DSV *Alvin* and the ROV *Medea-Jason*, owned by the Navy and operated by the Woods Hole Oceanographic Institution, are used full time for civilian science.

Some advanced research in vehicle-related technologies is being conducted at private oceanographic institutions in the United States, particularly by the Monterey Bay Aquarium Research Institute (in ROVs and AUVs) and by the Harbor Beach Oceanographic Institution (in DSVs).

UNDERSEA VEHICLE CAPABILITIES AND TECHNOLOGIES

As a first step toward setting development priorities, the committee reviewed each of the technologies for undersea vehicles at both the system and subsystem levels. Undersea vehicles should be considered as systems, in which the human operator and vehicle subsystems and components are integrated to achieve optimal performance. For that reason, any technology development program should focus on systems integration. Systems integration requires that vehicle subsystems, payload subsystems, and surface support be perceived as a whole.

Vehicle Subsystems

The committee reviewed each of the vehicle subsystems and made the following observations about their relationship to overall vehicle capability and their potential for improving vehicle performance.

- **Mission guidance and task control** together form one of the most fruitful areas of technology for undersea vehicles. Advances in the theory of automation have made possible so-called supervisory control, in which the human operator provides high-level, task-oriented commands rather than exercising direct control over all functions of the vehicle. The vehicle systems carry out the detailed control movements to accomplish the tasks. For example, when a sampling task is performed in mid-water, the moving vehicle and its manipulator are controlled as a single system. Further developments in this technology will have strong synergies with advances in navigation and sensors. Obviously, this technology area holds the greatest benefits for ROVs and AUVs. In the future, AUVs should be capable of pursuing tasks with abstract descriptions, such as finding and following a chemical gradient or surveying a given area with the ability to replan and reconfigure the mission based on a wide range of changing internal and external factors.
- **Communications** comprise a critical set of technologies for undersea vehicles, drawing on active developments in the electronics and telecommunications industries. DSVs use communications transmitted through the water acoustically; ROVs use copper or fiber-optic umbilicals; and AUVs communicate

through acoustic links. Recent improvements in digital acoustic communications, using data compression, show great promise for control of AUVs by using signals of increasing bandwidth for near real-time command. AUVs could use satellite communications to transmit data ashore using surface buoys equipped with transmitters (although bandwidth is limited by the instability of floating platforms). However, recovery to the surface of real-time data collected by payload sensors is still a major technical hurdle.

- **Data processing** holds many challenges, owing mainly to the need to manage the multiple streams of data from sensors of varied purposes, including both vehicle control sensors and payload sensors, such as sonars or video cameras. Modern database technology and automated processing of optical and acoustic images hold great promise for improvements in this area.
- **Navigation and positioning subsystems** use various combinations of acoustic sensors, video imaging, inertial guidance systems, and the Global Positioning System, depending on the application. The greatest advances in undersea navigation in the near future will come not from advances in particular systems but from integration of multiple subsystems and components.
- **Energy subsystem** improvement represents a variety of important development goals; the size, cost, and duration limitations of AUVs will be mitigated only when practical, safe, and readily available high energy density sources are developed. Energy sources have a lower development priority for DSVs and ROVs, the performance of which is limited by other factors.
- **Propulsion subsystems** represent mature technologies; existing systems are adequate for most foreseeable applications. The committee found no technology that promises significant improvements in the efficiency of propulsion.
- **Materials and structures** generally represent mature technologies, although some advances would be needed if deepest ocean-depth DSVs (capability to 11,000 meters) were sought. Also, some specialized materials, such as ultra-lightweight structural systems and anti-biofouling coatings, would be critical for long-duration missions.

Payload Subsystems

Payload subsystems are carried by vehicles for performing mission tasks. These subsystems involve the following elements:

- **Sensors** in the context of payload systems are used to collect data, as distinguished from those used in guiding and controlling the vehicle. They include various acoustic, optical, and chemical sensors; conductivity, temperature, and depth sensors; fluorometers and transmissometers; magnetic field sensors; gravity sensors; and current meters. All sensors have potential for important advances. However, improvements in acoustic and optical sensors should receive priority, owing to the breadth of their applications.
- **Task-performance subsystems** include manipulators and other tools. The development challenge is to take full advantage of new control techniques, with drills, cutting tools, wrenches, or other tools and with task-level control to enable operations like the gentle capture of a gelatinous creature. Research and development in the space program is an important potential source of advanced techniques.
- **Physical samplers** are used to collect samples in situ. An especially important development is physical samplers that have the ability to capture live organisms and subsequently maintain them at the same pressure and temperature.

Surface Support

Surface support includes logistics, positioning, data retrieval and processing, and launch and recovery. The launch, recovery, and vehicle handling functions are particularly important operations that reflect logistical support conditions, equipment capability, and crew training. Techniques for launching DSVs and ROVs are generally adequate; improvements are likely to have an incremental influence on overall system performance. An exception is research related to mitigating snap loads in long ROV umbilical lines when vehicles are deployed from small vessels; however, launch and recovery techniques for AUVs are still evolving. To take advantage of an AUV's lower cost and minimized surface support, new launch and recovery techniques will be required.

FOCUSING ON VITAL NATIONAL NEEDS

Assessing the subsystem technologies is insufficient without primary attention to full systems. To provide a grounding for this systems approach, the committee reviewed many of the ways undersea vehicles could help further the national interests in an extremely important range of scientific, regulatory, military nonclassified, and industrial applications.

Many scientific and technical fields are strongly dependent on undersea vehicles. Fundamental studies of deep ocean trenches, spreading centers, and faults using undersea vehicles have helped confirm and elaborate the theory of plate tectonics. Oceanographers can use these systems for studies of the geochemical and energy cycles of the atmosphere and ocean, bearing centrally on the future of climate and the biological productivity of the oceans.

Undersea vehicles are already showing evidence of their utility in observing commercial fish stocks and assessing harvesting techniques. They are critical to developing offshore oil and gas resources and laying undersea cable, as well as to the efforts of salvors and treasure hunters. Undersea vehicles may one day help in developing and monitoring waste storage sites on the remote deep abyssal plains at depths of more than 3,000 meters (approximately 10,000 feet).

To illustrate these potential contributions, the committee developed four "focal projects" describing potential applications of undersea vehicles in support of national needs (see Chapter 3). Some of these projects rely on technology that is available and economically feasible today; others are more visionary. All four projects address selected high-value missions or applications and employ multiple technologies useful in a wide range of undersea vehicle missions. They make possible tasks heretofore unachievable or impractical. They have potential for multiple use of vehicles or technologies for commercial, military, and scientific needs. Certainly they do not exhaust the potential for innovative uses of undersea vehicles. Nor are they intended to be a representative sample of potential applications. They simply offer a few challenging examples. In each case the committee established system performance requirements and commented on the state of the relevant technology.

All of the focal projects use AUVs as their primary vehicles. In the committee's judgment AUVs promise more payoff in advanced capabilities than DSVs or ROVs, which are technologically relatively mature. At the same time, DSVs and ROVs could be used in a variety of functions in these projects, not only in surveying and construction but also as alternatives to AUVs in some missions.

The focal projects for potential applications of undersea vehicles are as follows:

- **Synoptic Observation System**. This concept would use AUVs to make high-resolution measurements of chemical and temperature gradients in the water column over areas of several hundred square kilometers and for periods of up to several years. The AUVs would have a central control and battery-charging base and an array of bottom-fixed instruments and navigation aids. These vehicles will contribute importantly to modeling how the ocean functions to affect climate and the undersea food chain and to understanding tectonic processes (which relate to earthquake prediction).
- **Blue Water Oceanographic Sections and Hydrographic Survey**. These activities would involve the operation of one or more AUVs in synchrony with a surface ship to improve dramatically the accuracy and efficiency of oceanographic data gathering—another important part of modeling the ocean's function and predicting the climate.
- **Subsea Oil Field Inspection and Intervention**. This application of an undersea vehicle would use an AUV "garaged" at a central platform to make periodic inspections and maintenance of subsea oil and gas wellheads in the surrounding area—a task done better by an undersea vehicle, at much reduced cost, than by a surface-supported system. The subsea wellhead is an essential element in the development and operations of oil and gas resources in the new high-volume production fields of the Gulf of Mexico. In every case, the development of AUV capability to respond to shore-based, human, real-time task management will improve AUV operational cost efficiency by a major factor.
- **Search and Survey**. This effort involves the development of a versatile AUV system to carry varied suites of sensors, at depths of as much as 6,000 meters, to search the bottom for objects (such as lost aircraft) that have high value for the safety of air travel or regional defense and to make wide-area surveys of geological features (including mineral deposits) to assess resources of potential long-term value to the nation.

SETTING DEVELOPMENT PRIORITIES

The total system technology assessment and the review of national needs lead inevitably to the question of development priorities. This judgment involved a two-step analysis of the subsystem technologies. First, the committee ranked the subsystem technologies according to their impact on vehicle performance as "critical" (those whose improvement would enable important new capabilities, not otherwise achievable), "incremental" (those whose improvement offers significant, but not critical, improvements in efficiency or utility), or "mature" (those whose improvement would contribute only incrementally to vehicle performance). The committee then ranked the subsystem technologies deemed critical according to their likelihood of benefiting from development efforts. The technologies were ranked either "greatest potential" or "high potential." This sifting process revealed that three of the critical subsystem technologies offer the greatest potential for significant improvement in undersea vehicle performance and significant contributions to national needs:

- ocean sensors (acoustic, optical, and chemical)
- undersea communications (particularly digital acoustic methods)
- mission and task-performance control (with an emphasis on high-level, task-oriented control architecture)

These technologies should be the focus of investments in developing vehicle systems. However, major advancements in these three subsystem technologies will need to be integrated into complete vehicle systems, tested, and applied to missions such as the ones outlined in the four focal projects above.

ENHANCING THE NATIONAL CAPACITY

The national undersea vehicle capability for undersea work and research involves three aspects:

- technology development (by public and private sectors) necessary to carry out missions and meet identified national needs (see Chapters 2 and 4)
- availability and access to resources (coordination and scheduling of scientific and other uses) to scientific users with sufficient funding to support operations and studies (see Chapter 5)
- capital investment (by public and private sectors) in DSVs, ROVs, and AUVs and the support vessels and other assets that support their operation (see Chapter 4)

The nation currently lacks an effective system for coordinating and planning these capabilities. The government-operated systems that would support undersea science and ensure a national capacity for deep-ocean recovery and salvage are not well coordinated. Declining budgets for the Navy's submarine force, which provides many of the necessary systems, have put pressure on continued maintenance of the Navy's undersea vehicle assets; the users, who receive the benefits of these systems, have no authority or responsibility for their funding. The systems themselves are generally adequate for present purposes, but they are not efficiently managed, especially for science, and they will fall far short of meeting the critical national needs of the imminent future. The private sector has successfully managed investment in new technology and new systems of industrial interest, such as ROVs. These systems, adapted to scientific and national security missions, are increasingly capable and have important potential that is only beginning to be tapped.

A disciplined process of strategic planning is needed, with a long-term vision for assessing national needs and monitoring progress by the public and private sectors to meet those needs. Such a plan would not require a single, centrally directed undersea vehicle program; the vigor and variety of today's multiple-agency approach are valuable strengths. Coordination and planning can coexist with diversity, which is not to say that today's division of responsibilities must be maintained rigidly. Caught between declining military budgets and expanding ocean frontiers, the community must make better use of its resources and establish a more stable funding scheme.

CONCLUSIONS AND RECOMMENDATIONS

On the basis of its review, the committee developed the following conclusions with regard to undersea vehicles.

Conclusion 1. The nation has vital economic and scientific needs to significantly advance its capabilities for working, monitoring, and measuring in the ocean. Those needs involve national security, environmental protection, resource exploitation, and science. Undersea vehicles can contribute strongly to these capabilities by giving human beings access to new kinds of information about little known areas of the ocean and the seabed—information that may have a major impact on the well-being of large populations.

Conclusion 2. Technical advances are needed if the nation is to realize this potential; the priorities for these technologies are ranked in Chapter 4. The nation needs the ability to carry out construction support tasks, inspection, and maintenance in deep sea oil and gas fields safely and efficiently, using remote control. Autonomous undersea vehicles in synchrony with research vessels can help gather oceanographic and hydrographic data more accurately, quickly, and cheaply. Monitoring pollution and measuring the conditions that could lead to global climate changes will be easier with new chemical sensors. Surveying the bottom with the high resolution offered by undersea vehicles is likely to reveal valuable mineral deposits and assist in the location and recovery of objects related to public safety and security.

Conclusion 3. The committee finds the technological advances most critical to these important missions are in the areas of ocean sensors, subsea communications, and mission and task-performance control systems.

Conclusion 4. The vehicle technologies are generally mature enough to place the emphasis of technology advancement programs appropriately on systems integration.

Conclusion 5. Other countries today, like the United States in the past, have mounted focused programs with sustained support in the service of well-defined national needs.

Conclusion 6. The United States has no concerted program; instead it has a number of informally coordinated programs and no disciplined mechanism for long-term planning. The financial disjunction between users and the federal providers of undersea vehicles in some cases impedes coordination.

Conclusion 7. Failure to address the deficiencies of federal programs will constrain scientific progress, limit the nation's ability to develop and manage its ocean resources, and compromise national security and law enforcement.

The committee's recommendations are outlined below.

Recommendation 1. The nation should develop, maintain, and follow a long-term plan for federal undersea vehicle capabilities that takes into account all of the available facilities for undersea research.

Recommendations 2. In developing undersea technology the nation should meet its needs through combining government programs, joint technology agreements with foreign programs, and cooperative industry-government programs. Maximum use should be made of programs outside the federal government. All decisions should be based on the long-term plan recommended in this report.

Recommendation 3. Capital investment programs should take advantage of partnerships—from leases of U.S.-certifiable foreign submersibles to joint development and use of new vehicles and support vessels with industry and foreign programs. The federal government should buy wholly new vehicles for civilian use only when other sources are not available and the national interest (as determined by the planning process recommended here) demands it.

Recommendation 4. In ensuring user access to undersea vehicles, the nation should maintain the pluralism of the present approach with a variety of funding agencies. The flexibility and local innovation of that approach are major strengths of the U.S. system of science and technology. At the same time, the agencies involved should be guided by a shared strategic view of future needs.

Recommendation 5. Stable funding should be provided for those undersea vehicle systems that are viewed as national assets.

1

Introduction

The United States, by most standards, is the foremost maritime trading nation of the world. Nearly one-fourth of the world's trade flows through U.S. seaports. The U.S. Navy is the largest standing navy in the world. Investments by U.S. industry and government in ocean science and technology are unparalleled. In addition, U.S. research in the ocean sciences currently leads the world in most disciplines, from physical oceanography to marine biotechnology, and the U.S. educational establishment is a world leader in ocean science and engineering. But support for that research has declined steadily in the United States, while the nation's need for knowledge about the oceans has grown.

The national importance of maritime activities—both for managing resources and augmenting scientific understanding—will continue to increase in the decades to come. Studies of the chemistry and energy flows of the ocean and atmosphere are critical for predicting how the Earth's climate will change. Pollution from human activities needs to be assessed accurately so that appropriate action may be taken. Ocean plant and animal resources require better management to help feed the world's swelling population. (Stocks of fish, in particular, are increasingly heavily exploited throughout the world and need better monitoring and ecological study.) Nonliving resources such as minerals and various forms of energy, both renewable and nonrenewable, could be supplements to terrestrial supplies. Scientific studies of the Earth's crustal plate boundaries (fundamental features of the planet's structure, most of which are underwater) will be needed to help predict earthquakes, among other reasons. The stunning discovery of hydrothermal vents at the mid-ocean plate boundaries and the associated food chains based not on photosynthesis, but on chemosynthesis, is another example of the rich contributions that ocean exploration offers.

In recognition of the growing importance of the oceans to human society, the United Nations Law of the Sea Treaty has come into force. This treaty spells out national rights and responsibilities for exploiting and protecting the ocean environment. For the United States, the agreement added responsibilities for administering the use of the ocean surface, associated seafloor, and subseafloor in an area equivalent to 75 percent of the nation's land area. The new realm is the exclusive economic zone (EEZ), a roughly 200-mile band along the coastlines over which the United States has exclusive development rights and responsibilities for environmental protection.[1] The treaty will establish a more secure environment for investment, encouraging activities for managing and exploiting ocean resources.

Oceanographers, ocean engineers, and those who use and manage the ocean's resources use a variety of platforms and payloads, including Earth-orbiting spacecraft, surface ships, moored and drifting buoys, towed sleds, and undersea vehicles, depending on the task at hand. To meet our national needs and commitments, these scientists, engineers, and managers will need to enhance the capabilities and productivity of methods for making measurements under the sea and on the seafloor and for performing work in sectors of the sea that are difficult, if not impossible, to access. This report focuses on undersea vehicles, which may serve as platforms for varied sensors and work devices and may be operated as independent units or as parts of larger systems, such as ships, satellites, buoys, and towed sleds. It should be noted that many technical issues related to sensors, communications, navigation, and power are common to undersea vehicles, buoys, and sleds.

Undersea vehicles have replaced divers in many applications because they can go deeper, stay down longer in greater safety, and carry a variety of sensors. The evolving technologies of remote sensing, remote control, and automation ensure that undersea vehicles will continue to extend the

---

[1] In 1994, U.S. representatives signed an international agreement governing seabed minerals beyond the 200-mile limit under the International Convention on the Law of the Sea. By doing so, a sequence of negotiations going back several decades was completed. The convention, signed in 1982, had left successive U.S. presidential administrations dissatisfied with the seabed minerals provisions. The 1994 agreement allayed these objections. In practical terms the agreement opened U.S. firms' access to these seabed minerals for the first time (subject to ratification by the Senate).

human reach to the deepest parts of the ocean. Some of the functions performed by undersea vehicles are:

- making observations, either with remote sensors (e.g., video or still cameras, laser scanners, or scanning sonars) or through direct observation by occupants of vehicles
- logging or displaying data from contact sensors (e.g., physical, chemical or biological instruments)
- manipulating tools or work packages to perform scientific and industrial tasks
- retrieving samples or artifacts
- navigating beneath the surface, logging the vehicle's trajectory
- responding to control instructions from a human operator or computer

Many of these capabilities are in use or under development in industry to support the offshore oil and gas industry and other marine industries. Others contribute to government missions and will require development support from government. The remaining chapters of this report detail current vehicle technical capabilities, present and potential applications for vehicles, needed technical advances, the committee's recommendations for priorities in development and how to achieve and implement these improvements.

UNDERSEA VEHICLES DEFINED

An undersea vehicle is a mobile, controlled, self-propelled subsurface platform capable of carrying sensors and tools. It may or may not be occupied and/or piloted by humans. Three broad types can be distinguished: deep submersible vehicles (DSVs), remotely operated vehicles (ROVs), and autonomous underwater (or undersea) vehicles (AUVs). While these broad categories are useful for analysis and discussion, there are many variations and hybrids. For example, an untethered vehicle with high-level control provided by an acoustic data link to a supporting surface ship or shore station is not completely autonomous; nevertheless, it is classified as an AUV in this report.

Undersea vehicles are not new, having functioned in some form or another for over four decades. However, the rapid advances in technologies and systems integration that have taken place in the past two decades have allowed great improvements in the performance of sensors, communications, navigation, and control. New capabilities and scientific uses for submersibles have emerged, and others are on the horizon that could greatly expand the type, quantity, and quality of information that can be derived. Because of these improvements, use of undersea vehicles is being considered in disciplines where they have seen very little activity in the past, such as physical and chemical oceanography, as well as in marine geology, biology, and mid-water benthic studies, where they are currently used. Many of these same improvements have also enhanced the performance of buoys and towed sleds.

Deep Submersible Vehicles

More than 200 DSVs have been built worldwide since the late 1950s. DSVs are human occupied, generally by a pilot and a crew of two. Most are deep probes that must travel more vertical than horizontal distances to accomplish their missions. DSVs carry tools, sensors, and sampling devices; and missions generally do not exceed 16 hours. DSV diving operations are supported by mother ships. Most DSV operations involve many dives without returning to port. Operating depths range from a few hundred meters to nearly 6,500 meters. The committee estimates that today about 42 DSVs are operational worldwide. That estimate is uncertain, however, because some of the activity in this field has been carried out by the military and is classified. Appendix E lists the world's known DSVs. Figure 1-1 shows *Alvin*, which is the oldest DSV in service and has completed the greatest number of dives; Figure 1-2 illustrates *Deep Rover*, one of the newest DSVs. There are also more than 40 tourist submersibles in service throughout the world that carry up to 64 passengers each on rides to depths of 30 to 50 meters. These submersibles share some attributes with DSVs. They help maintain the industrial base and development of some materials and provide useful operational training and maintenance experience. However, this committee does not consider them as being DSVs since they are not designed or intended to perform work tasks and because of their shallow depths of operation (< 50 meters).

Remotely Operated Vehicles

ROVs are unoccupied, tethered submersibles with umbilical cables that carry electrical power, sensor data, and control commands. The pilot or operator is remote from the work site but maintains control of the vehicle. ROVs are used mainly in offshore oil and gas operations for various inspection and manipulation tasks, and they have replaced divers in many jobs in that industry. ROVs are also widely used for laying undersea cables, and, as offshore development moves into deeper waters, ROVs will be increasingly required.

Typically, the human operator of an ROV works at the telemanipulation level, piloting the vehicle and controlling its manipulators through direct coordinated control using joy sticks. However, recent advances in control architecture provide the option of using object-oriented, task-level control, which allows the operator to issue commands at the task level rather than having to coordinate vehicle manipulator and instrument control operations continually. This innovation allows the operator to return to the lower level task when necessary. Vehicle sizes range from small, portable, swimming television cameras that weigh only a few kilograms to

FIGURE 1-1 Photo of *Alvin* (Woods Hole Oceanographic Institution/Navy). The U.S. Navy-owned *Alvin* has been operated by the Woods Hole Oceanographic Institution since 1964. Its depth capability is 4,500 meters. It carries a crew of three and has made over 3,000 scientific dives. Source: Woods Hole Oceanographic Institution.

FIGURE 1-2 Photo of *Deep Rover* (Deep Ocean Engineering, Inc.). Two *Deep Rover** 1002s were built by Deep Ocean Engineering, San Leandro, California, in 1994 and delivered to a French television production company. These are two-person DSVs, designed to operate to 1,000 meters. Source: Deep Ocean Engineering, Inc. *Registered trademark of Deep Ocean Engineering, Inc.

huge work systems that can weigh several metric tons. The committee estimates that since the early 1970s, more than 1,000 ROVs have been put into service worldwide. Depth ranges for ROVs are now from a few meters to 11,000 meters. Figures 1-3, 1-4, and 1-5 illustrate a cross-section of ROV types: *Phantom* (Deep Ocean Engineering), a light vehicle for inspection and observation tasks; *Triton 19* (Perry Tritech), a heavy-work vehicle typically used in the offshore oil industry; and *Ventana* (Monterey Bay Aquarium Research Institute), designed for science missions.

Autonomous Underwater Vehicles

AUVs are unoccupied submersibles, generally without tethers or umbilical cables, although some are fitted for specific missions with a very fine fiber-optic filament that can transmit data to the launching station. All power is supplied by onboard energy systems (such as batteries or fuel cells). Nearly all AUVs are developmental systems. AUVs encompass a range of systems, from relatively simple and inexpensive to quite complex. They share many key technologies (sensors, power, navigation, and communications) with moored and drifting buoys.

AUVs today typically have entire missions programmed in their onboard controllers before launch. Remote control

FIGURE 1-3 Photo of *Phantom* (Deep Ocean Engineering, Inc.). The *Phantom** HD2+2 was developed by Deep Ocean Engineering, San Leandro, California, in 1985. There are 14 *Phantom* designs in the series. The Deep Ocean Engineering, Inc. has delivered over 300 *Phantoms* to 22 nations and 7 navies. Source: Deep Ocean Engineering, Inc. *Registered trademark of Deep Ocean Engineering, Inc.

FIGURE 1-4 Photo of *Triton 19* (Perry Tritech). The Perry Tritech *Triton 19* is designed for operations to 1,000 meters deep and weighs 2,450 kilograms in air. *Triton 19*, which has a 227-kilogram payload capability, began operations in 1994 in support of the Shell Auger Tension Leg Platform in the Gulf of Mexico. Source: Perry Tritech, Inc.

communication is not practical without a cable to carry signals because radio waves are strongly attenuated in water. However, advances in acoustic communications have increased the effective bandwidth, and new architectures for task-level control, which has a lower bandwidth requirement than low-level control, have made real-time human control of untethered vehicle missions a powerful new possibility. (Of course, incorporation of such systems would blur the distinction between autonomous and remotely controlled categories of vehicles.) The military has supported most of the developmental work on AUVs. Fewer than 100 AUVs have been built worldwide since 1970, nearly all of

FIGURE 1-5 Photo of *Ventana* (Monterey Bay Aquarium Research Institute). This vehicle conducts a variety of oceanographic research tasks in biology, geology, and chemistry, to depths of 1,850 meters. Operational since 1988, it has made over 900 research dives and has logged more than 5,000 hours under water. Source: Bruce H. Robison, Monterey Bay Aquarium Research Institute.

FIGURE 1-6 Photo of *UUV II* (Charles Stark Draper Laboratory for DARPA). The DARPA-U.S. Navy testbed unmanned, undersea vehicle was designed and built by the Charles Stark Draper Laboratory. Two vehicles have been built and are intended to validate enabling technologies for the Navy's tactical acoustical system and the autonomous mine-hunting and mapping mission. Source: Charles Stark Draper Laboratory, Cambridge, Massachusetts.

them by the military. Small, low-cost AUVs, such as the *Odyssey* system being developed by the Sea Grant Undersea Vehicles Laboratory at the Massachusetts Institute of Technology (MIT), offer the potential for use in multiples. If costs can be kept low enough without compromising performance, they could be deployed in fleets to make observations at high temporal and spatial resolution (see, for example, Curtin et al., 1993, and Stommel, 1989). Figures 1-6 through 1-9 illustrate several AUVs, showing a range of vehicle sizes.

FIGURE 1-7 Photo of the autonomous benthic explorer (*ABE*) (Woods Hole Oceanographic Institution). *ABE* was designed and built by Woods Hole Oceanographic Institution with National Science Foundation funding to monitor deep ocean hydrothermal systems. *ABE* can operate to a depth of 6,000 meters and remain on station for many months. Source: Woods Hole Oceanographic Institution.

FIGURE 1-8 Photo of *Theseus* (International Submarine Engineering, Ltd.). *Theseus* is a multipurpose autonomous underwater vehicle developed by International Submarine Engineering, Ltd. for the Canadian Defence Research Establishment Pacific. The vehicle is undergoing test trials. Operations to 1,000-meter depth extending up to 350 kilometers have been demonstrated. Source: International Submarine Engineering, Ltd.

FIGURE 1-9 Photo of *Odyssey I* (Massachusetts Institute of Technology). *Odyssey I*, which was launched in 1992, is the first in a series of small AUVs designed and operated by the Sea Grant Undersea Vehicles Laboratory to map under ice fields and respond to volcanic events near ocean ridges. Operational depth is 6,000 meters, and displacement is 195 kilograms. Source: Massachusetts Institute of Technology Sea Grant.

EVOLUTION OF UNDERSEA VEHICLES SINCE THE 1950s

In the United States, as in most other maritime nations, interest in the exploration of the sea has come in waves, each prompted by an industrial or governmental need. With each wave has come a new generation of undersea vehicles. Undersea vehicles made their first important contributions before World Ward II (in gravity measurements for the U.S. government), but their serious development in the United States did not begin until the 1950s.

Vehicle Development from the 1950s through the 1990s

The 1950s. In the 1950s, the cold war rivalry with the Soviet Union led to the first nuclear submarine; a submerged circumnavigation of the world; and, in 1959, the first transits under the North Pole by the U.S. Navy nuclear submarines *Nautilus* and *Skate*. In 1958, the Navy also purchased the world's deepest diving submersible, the bathyscaphe *Trieste*, which had been developed by private interests in Switzerland and Italy.

The 1960s. The 1960s saw a continued military emphasis on methods for performing undersea tasks, but it was submersibles, human occupied and unoccupied, and advanced diving technology that were prominent at this time. In 1960, the Navy's *Trieste* successfully dove to the deepest place in the world ocean (the Challenger Deep in the Mariana Trench, approximately 11,000 meters deep) in the culmination of a year-long deep diving project. The bathyscaphe *Trieste* was a design first built in the 1940s; later versions of the *Trieste* were operated until 1982. Navy laboratories and government contractors also developed the first practical ROVs in the 1960s. The ROVs were intended for use in recovering torpedoes from the seafloor. The controlled underwater recovery vehicle (CURV) series of ROVs emerged from this development, and a CURV was used to recover the nuclear bomb from the sea off Palomares, Spain, in 1966. Also during the 1960s, the Navy's Sea Lab pursued a program of human habitation on the seafloor that pushed the limits of diving physiology and technology. The 1960s also found many of the large aerospace companies entering the submersibles field, and many new human-occupied vehicles were built and put into service by companies such as Grumman, Lockheed, General Motors, General Dynamics, and Westinghouse. Few of these vehicles remained in use beyond the end of the decade.

The 1970s. In the 1970s, development of undersea vehicles was split between the defense and civilian sectors. Military interest centered around the requirements of antisubmarine warfare (ASW). Vehicles were needed, for example, to gather oceanographic data, to act as decoys, and to carry out searches. Civilian programs supported offshore oil and gas development, which moved increasingly into deeper waters and included exploring for seafloor hydrocarbons and supporting production operations in the North Sea. In the offshore industry DSVs and ROVs replaced divers in more and more of the areas of marine structure construction, inspection, and servicing.

By the end of the 1970s, virtually all of the DSVs had been replaced by ROVs. A significant exception was a handful of DSVs used mainly for marine scientific research. In the United States, examples included the Navy-owned *Alvin* (operated by Woods Hole Oceanographic Institution), *Sea Cliff*, and *Turtle*; the *Johnson Sea-Link*, owned by the Harbor Branch Oceanographic Institution; and the *Delta*, owned and operated by a for-profit company that primarily supplies oceanographic research services. These vehicles have made possible important scientific discoveries. *Alvin*, for example, provided access to the hydrothermal vents at the ocean spreading centers. The vehicles are also used for searches of the bottom and other work tasks. As shown in Figure 1-10, *Johnson Sea-Link* is configured for civilian scientific missions.

FIGURE 1-10 Photo of *Johnson Sea-Link* (Harbor Branch Oceanographic Institution). The *Johnson Sea-Link I* and *II* DSVs were designed and built by Harbor Branch Oceanographic Institution and commissioned in 1971 and 1972. These vehicles are intended for research in the marine sciences, are classed and certified to a maximum operating depth of 2,344 meters, and have conducted over 7,000 dives. Source: Harbor Branch Oceanographic Institution.

During the 1970s, significant AUV development began. The University of Washington's two AUVs, *Spurv* and *Uars*, successfully collected oceanographic data, including some from under the ice. The first deep-diving AUV, France's *Epaulard*, performed more than 500 dives—many of them to 6,000 meters—operated video cameras, and communicated with the surface over an acoustic data link.

The 1980s. In the 1980s, concern with ASW continued. Growing Navy spending supported the development of several early AUVs for various military missions. The number of piloted submersibles available worldwide continued to decline as commercial users shifted their interests to ROVs, which were demonstrating increasing work capabilities at lower cost each year and relieved concerns about risks to human operators. By the early part of the decade, ROVs became the dominant platform for performing undersea work offshore. In addition to the work-class ROV, the early 1980s witnessed the emergence of a large variety of small (<50 kg) ROVs, sometimes called low-cost ROVs, or LCROVs, many of which were designed for special purposes or missions.

To meet scientific and defense-related objectives in the 1980s, four nations (France, Japan, Russia, and the United States) organized, built, and operated DSVs that could be used at depths to 6,500 meters. These vehicles remain operational today. The U.S. Navy *Sea Cliff*, modified from a shallower-diving DSV, is one of five deeper-diving DSVs developed in the 1980s. France's IFREMER[2] designed and built the 6,000-meter *Nautile*, and the Russian Academy of Sciences purchased two Finnish-built *MIR*-class submersibles that also can operate to that depth. The Japanese *Shinkai 6500* (see Figure 1-11), launched in 1989, can dive to 6,500 meters; it is the deepest-diving DSV in the world today.

The few AUVs built in the 1980s were mainly experimental vehicles. The goal of Unmanned Undersea Vehicle

---

[2]Institut Français de Recherche pour l'Exploitation de la Mer.

FIGURE 1-11 Photo of *Shinkai* (Mitsubishi Heavy Industries/Japan Marine Science and Technology Center). *Shinkai 6500* was built by Mitsubishi Heavy Industries, launched in 1989, and is operated by Japan Marine Science and Technology Center. *Shinkai 6500* carries a crew of three and is capable of diving to 6,500 meters, making it the deepest diving DSV in the world. Source: Japan Marine Science and Technology Center.

project, begun in 1988 under a joint Defense Advanced Research Agency (DARPA)[3] U.S. Navy Program was to demonstrate that AUVs could fulfill special Navy mission requirements. The two testbed vehicles that emerged from this project were designed and built by the Charles Stark Draper Laboratory (see Figure 1-6). Since the inception of the program, the goals have been modified to include technology development critical to increasing the role of AUVs in undersea warfare, particularly in minehunting and mapping. The project focus is on fuel cells, acoustic communications, and more accurate navigation systems. The two ARPA-Navy vehicles are large,[4] medium-depth-capable vehicles that continue to serve as testbeds. Classified AUVs may still be operated by the military, but such information was not available to the committee.

The 1990s. The increasing utility and reliability of ROVs in offshore operations during the 1980s resulted in broad acceptance of this type of vehicle platform, including for applications outside of the industry, such as underwater telecommunications and nuclear facility inspection and maintenance. By the early 1990s, both ROVs and their services were commonly supplied by offshore oil service organizations rather than by oil companies.

One type of ROV range and complexity is illustrated by *Kaiko* (see Figure 1-12), which was developed and tested by the Japan Marine Science and Technology Center (JAMSTEC). In March 1995, *Kaiko* dove to the deepest known place in the world ocean, the Challenger Deep, in the Mariana Trench in the Pacific Ocean (11,033 meters).

Public interest in undersea vehicles has been captured by a series of deep ocean explorations in the past decade. Using submersible platforms has facilitated major deep-seafloor discoveries in the life sciences and geosciences and has provided dramatic video of towering deep-ocean vents in action. But it has been the discovery and imaging of famous shipwrecks, such as the *Titanic*, *Bismarck*, and *Lusitania*, by teams from the Woods Hole Oceanographic Institution, led by Dr. Robert Ballard, that have made people worldwide aware of what can be done in the deep-ocean with in situ platforms. In the United States, the privately funded Jason Project uses telecommunications to create a powerful educational tool that lets students observe deep-diving expeditions in real time, via satellite, and even lets them take control of the remote manipulators. Recovering evidence from crashed aircraft, for example, flight recorders, doors, and engines, at depths up to 6,000 meters has made deep-ocean search and recovery seem routine to much of the public. However, the

---

[3]For a long time called DARPA, the agency was renamed ARPA and has now reverted to the original name. Both terms are used in this report depending on the related time frame or publication.

[4]These unmanned undersea vehicles are 11 meters long, 1.1 meters in diameter, and weigh 6,800 kilograms. The are capable of diving to 304 meters (*UUV 1*) and 457 meters (*UUV 2*).

INTRODUCTION

FIGURE 1-12 Photo of *Kaiko*. Built by Mitsui Engineering and Shipbuilding Company and operated by Japan Marine Science and Technology Center, *Kaiko* reached the deepest ocean depth at approximately 11,000 meters in March 1995, establishing a record depth for an undersea vehicle (non-bathyscaphe). The system includes two elements shown in this photo: the lower part is the ROV itself; the upper part is its "garage" or carrier. Source: Japan Marine Science and Technology Center.

search and recovery capabilities are quite limited, and the deepest ocean areas—beyond 6,000 meters, down to 11,033 meters—remain out of practical reach.[5]

U.S. TRENDS IN UNDERSEA VEHICLE DEVELOPMENT AND USE

U.S. government and industry investment in undersea vehicle research and development today is far less than during the peak years, the 1960s and 1970s. The U.S. Navy, a primary sponsor of underwater vehicle research and development, achieved most of its fundamental technical objectives for undersea vehicles in the 1960s and 1970s. Those DSV assets that remained, mainly in the U.S. Navy, are substantial, although increasingly outdated. In the civil sector, the initial rush of aerospace companies into DSV development ended, and the offshore oil and gas industries began to find uses for ROVs. In the 1980s, the decline in military development of DSVs was not offset by a pickup in commercial demand. These factors resulted in considerable cutbacks in fast-moving, innovative technology development.

---

[5]Only about 2 percent of the ocean floor lies deeper than 6,000 meters.

With the end of the Cold War, the Navy opened access to its DSVs more widely for civilian scientific research. The Deep Submergence Science Committee (DESSC) provides schedule coordination and oversight for many of the DSVs supporting this research, including *Alvin* and more recently *Jason-Medea*. Supported jointly by the National Oceanic and Atmospheric Administration (NOAA), the National Science Foundation (NSF), and the Navy, DESSC is supported by a University-National Oceanographic Laboratory System grant funded by the National Science Foundation, the National Oceanic and Atmospheric Administration, the Office of Naval Research (ONR), and other agencies. The U.S. Navy DSVs (*Turtle* and *Sea Cliff*), are now used to support civilian-related research up to 60 days each year. *Alvin*, owned by the Navy and operated by the Woods Hole Oceanographic Institution, is used for civilian science projects.

The National Undersea Research Program (NURP) of NOAA operates and leases undersea vehicles, making them available for scientific missions. Available DSVs include the *Johnson Sea-Link I* and *II* and *Clelia*, owned by the Harbor Branch Oceanographic Institution, and the *Delta*, owned by Delta Oceanographics. A variety of smaller ROVs, owned by NURP centers, are also available for scientific missions.

The last government-financed DSVs were completed a quarter of a century ago—two highly specialized deep submergence rescue vehicles (DSRVs). In the commercial sector only four work submersibles have been completed since 1983. Similarly, U.S. government support for developing ROVs virtually ceased 15 years ago, except for the Navy-supported *Jason-Medea* ROV project (>2,000-meter depth), which gave scientists an ROV able to reach deep waters. The offshore oil and gas industry is the primary motivating force in ROV technological advances. Most government procurements for these vehicles are for modified off-the-shelf equipment from commercial vehicle makers. Makers of AUVs are receiving modest amounts of research and development funding from the U.S. government or are developing these vehicles in-house. The committee was unable to obtain budget figures for government programs.

Efforts are being made to transfer more military-developed technology (including undersea vehicle technology) to the civil sector. Some vehicles, and their related technologies, that were originally developed for national defense are now available for transition to dual military and civil use. The Naval Command, Control, and Ocean Surveillance Center in San Diego, California, for example, recently turned the autonomous unmanned search system (*AUSS*) over to Oceaneering International, a contractor for the Navy Supervisor of Salvage. This AUV previously was used for classified search missions.

The basic technologies used to build AUVs are largely unclassified, but the committee did not have access to classified government work. Military research and development of undersea vehicles may be in progress, and military vehicle systems may be operational; these systems may already have capabilities that are not yet developed in the civil sector. Since the civilian sector was the focus of this study, the committee could include only unclassified military technologies that could be transferred from defense programs.

FOREIGN PROGRAMS

The U.S. government, which pioneered undersea vehicle technology in the 1950s and 1960s, cannot be said to place a high priority on undersea vehicles today. As noted in the previous section of this chapter, the government has gradually withdrawn support for its DSV programs, beyond what is necessary to keep them running. The U.S. Department of Defense supports some AUV research and development, but most ROV activity remains in the private sector.

Other nations have assumed leadership in many areas of undersea vehicle technology, although no single nation has assumed broad world leadership. Through JAMSTEC, Japan has pursued an intensive national program of undersea exploration using undersea vehicles, including an ambitious program of diving to nearly 11,000 meters—the very bottom of the ocean—with the ROV *Kaiko* (working with the 6,500-meter, tether DSV, *Shinkai*). The general goal of the Japanese program is to understand the Earth's crustal structure and seabed resources.

Some European governments (e.g., the United Kingdom, France, and Norway) also tend to support research and development in this area more generously than does the United States. The economic stimuli in Western Europe are largely the need to manage fisheries and develop offshore oil and gas sources. The European Commission funds cooperative research and development in subsea technology, including the development of standards (Seymour et al., 1994). Commission funding of cooperative, multinational projects in vehicle automation, sensing, and manipulation through the Marine Science and Technology (MAST) program is also significant—on the order of 25 million ECU annually (about $31 million).

The 20 or more undersea vehicles developed by the former Soviet Union are largely the products of military programs. Today they are carried on in the hopes of marketing the military technology. Both Russia and Ukraine have skilled work forces and sophisticated test facilities and have produced an array of DSVs, ROVs, and AUVs, including impressive ROVs and AUVs developed in the Russian Far East, and the *MIR*-class DSVs (Mooney et al., 1996). A range of advanced hull materials, some derived from spacecraft, are a particular strength of these programs. As of mid-1994, the Russian and Ukrainian programs were inactive except for the vehicles under contract to Western users; however, both countries still have important potential as research and development centers (Seymour et al., 1994).

France's program, run by IFREMER, emphasizes DSVs, notably the 6,000-meter *Nautile*. However, IFREMER also has an ambitious 6,000-meter, science-dedicated ROV development program. French programs leading to integration of local sensor data for navigation and control also have great potential (Seymour et al., 1994).

The United Kingdom emphasizes the development of advanced sensors and affordable AUVs and ROVs that can be used in research and in the offshore oil and gas industry (Seymour et al., 1994). The United Kingdom initiated what is now known as the Autosub Program. The objectives of this program are to build and test a proof-of-concept vehicle and construct two AUVs: *Dolphin*, which would be capable of crossing the north Atlantic Ocean, taking soundings and samples and transmitting data by satellite during periodic surface excursions; and *Doggie*, which would cover the ocean floor and acquire high-resolution data from high-frequency instruments. *Doggie* would support a subbottom profiler, magnetometer, and chemical sensors. At present, Autosub focuses on technology development rather than vehicle integration.

Norway, like the United Kingdom, is faced with the need to monitor and inspect deep-water pipelines in the North Sea. Norway also places a high priority on monitoring its waters for contamination from a number of scuttled Soviet nuclear submarines in the Arctic Ocean.

Canada's long coastline, some of it icebound during all but two or three months of the year, has induced the Canadian armed forces to support development of undersea platforms to access areas beyond the reach of surface support. Moreover, the ice-free fiords in the Vancouver, British Columbia, area offer excellent vehicle test sites adjacent to centers of research. Canadian development is continuing, with emphasis on AUV applications.

FINDINGS

Finding. The oceans are regions of vital human concern. Their study is necessary to understand the future of the global environment, sustain exploitation of living and nonliving resources, and advance human knowledge. The United States investment of human and capital resources in marine science and technology is probably the largest worldwide.

Finding. Undersea vehicles are important components of the nation's capability to work and study in the oceans. In combination with satellite remote sensing, surface vessels, and a variety of towed sleds and moored and drifting buoys, undersea vehicles offer unique qualities: flexibility, control, and direct human access to the deep.

Finding. From the 1960s to the end of the 1970s, the United States was a world leader in the development of all undersea vehicles; however, since that time much of this leadership has been lost. The major exceptions are U.S. development of ROVs for commercial use in the 1980s and some limited forms of AUV research supported by the military.

REFERENCES

Curtin, T.B., J.G. Bellingham, J. Catipovic, and D. Webb. 1993. Autonomous oceanographic sampling networks. Oceanography 6(3):86–94.

Mooney, J.B., H. Ali, R. Blidberg, M.J. DeHaemer, L.L. Gentry, J. Moniz, and D. Walsh. 1996. World Technology Evaluation Center Program. World Technology Evaluation Center Panel Report on Submersibles and Marine Technologies in Russia's Far East and Siberia, in press. Baltimore, Maryland: Loyola College of Maryland, International Technology Research Institute.

NRC (National Research Council). 1992. Working Together in the EEZ: Final Report of the Committee on Exclusive Economic Zone Information Needs. Washington, D.C.: National Academy Press.

Seymour, R.J., D.R. Blidberg, C.P. Brancart, L.L. Gentry, A.N. Kalvaitis, M.L. Lee, J.B. Mooney, and D. Walsh. 1994. World Technology Evaluation Center Program. Pp. 150–262 in World Technology Evaluation Center Panel Report on Research Submersibles and Undersea Technologies. Baltimore, Maryland: Loyola College of Maryland.

Stommel, H. 1989. The Slocum Mission. Oceanus 32(Winter 89/90): 93–96.

2

Undersea Vehicle Capabilities and Technologies

This chapter describes the types and capabilities of undersea vehicles that will provide the functions necessary to serve as undersea tools for science and industry and respond to vital national needs in the oceans. When describing its capability, it is essential to conceive of an undersea vehicle as a system that encompasses: (1) a vehicle platform plus vessel and shore-based support; (2) a payload of task-performing devices and sensors; (3) a human in control, located either remotely or on site; and (4) the technology and hardware necessary to support, launch, and retrieve the vehicle. This system-level vehicle concept must take explicit account of potential missions and objectives.

This chapter points out the importance of system integration—the "glue" binding the human-vehicle system into a single operational unit. The contribution of each subsystem to the overall technical performance capability of the vehicle assembly is assessed in the system context. Advances in the technology and capabilities of some subsystems may provide more overall system performance benefits than would improvements in other subsystems. Finally, and very importantly, each key vehicle subsystem is assessed in regard to the current state of practice and where significant improvements can be made.

In this systems context, an assessment of the state of practice is the starting point for assessing vehicle technology and capability for responding to the nation's needs. An understanding of development trends and influences within, and external to, the undersea vehicle industry is the next important step in determining research and development strategies and priorities. Indeed, some of the valuable aspects of vehicle development are driven largely by advances in other industries, both domestic and foreign. This relationship is a natural process induced by the intense competition among a few small vehicle engineering and manufacturing organizations that must focus their limited development capital on adapting technologies to the special requirements of the undersea environment and missions. The committee identified those technologies that are vital to advances in undersea vehicle capability and are not already being pursued in other arenas.

VEHICLE SYSTEMS

Vehicle systems designs are driven by mission requirements, available technology, and cost. A system must be an effective combination of human operators and vehicle subsystems and components that are integrated to achieve optimal performance (see system elements schematic shown in Figure 2-1). Consequently, the task for builders and designers of a vehicle system is to integrate this multiplicity of subsystems into a working whole. The attributes of each subsystem will be assessed and established by the committee based on the state of technology, relative cost, and mission requirements.

For a given mission or task, designers and engineers begin by considering its particular requirements. These vary according to mission requirements for maximum depth, endurance in range and in time, and sensors; the optimal human role in command and control; power available; cost; and a host of other factors. Depending on these criteria, a particular type of vehicle (DSV, ROV, or AUV) is selected and its subsystems chosen to best fulfill mission requirements. Subsystem capabilities improve with time as technology develops, so the optimum solution in the near future may be significantly better than what is optimal using today's technology.

Table 2-1 summarizes typical capabilities and examples of DSVs, ROVs, and AUVs. The characteristics listed are generic, and there are exceptions; but the table outlines the characteristics and relative strengths and limitations inherent to each type of system.

DSVs place the human in the environment and benefit from the human's high-resolution, three-dimensional observation capability and full visual depth of field that is still superior to the observational capabilities provided by remote sensors. This capability enhances the performance of inspection and imaging tasks as well as manipulation. In general, today's DSVs have relatively large payload capacities and good manipulation capabilities but do not provide real-time feedback to the surface. Moreover, they require costly

CAPABILITIES AND TECHNOLOGIES

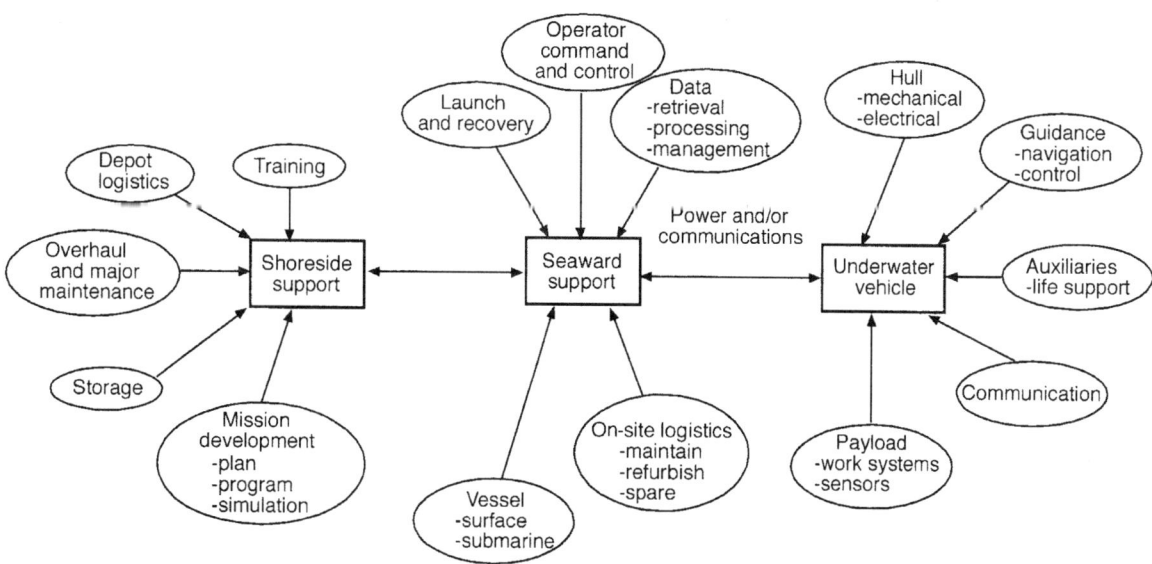

FIGURE 2-1 Schematic diagram of vehicle systems.

pressure housings and life support and safety systems for the human operators. DSV endurance at the work site is limited by prolonged surface periods for crew change and vehicle replenishment. The cost per operating hour is usually much greater than for other vehicle types because of the extra cost attributable to crew accommodation and life support, the larger surface support ship needed to handle the heavier human-occupied vehicle in launch and retrieval, and the more limited availability—therefore, the normally higher day rate—of support ships. In addition, the capital cost component of the day rate for DSVs is potentially higher than for unoccupied vehicles because they can be operated fewer hours due to human limitations.

In comparison to DSVs, ROVs provide greater endurance and greater range, including maneuverability of surface support vessels, at lower cost. Because life-safety support is not required, ROVs can operate in hazardous environments and provide simultaneous real-time observation and control to multiple remotely situated observers (Robison et al., 1992; Bowen and Walden, 1993).

AUVs are free from the tether restraint common to ROVs and can perform tasks with little or no operator input. However, AUVs must operate on a limited energy budget and can provide little real-time feedback to the operator, who is limited by the bandwidth available when using acoustic communication. However, new advances in task-level-control architecture will enable the human to command tasks at high level in real time, and acoustical advances are continually expanding the available acoustic-communication bandwidth.

DEEP SUBMERSIBLE VEHICLES

In the 1960s and 1970s many DSVs (with a human in situ) were in operation in a variety of different applications worldwide. Estimates indicate that more than 100 DSVs of all types were built during this time. DSVs have dominated ocean exploration, and some systems, such as *Alvin*, have been operational for the last three decades. DSVs continue to be used in support of certain military tasks and marine scientific research. However, few have been built recently (only four since 1990), and the number operating today is significantly less than it was 20 years ago. Since the 1970s, ROVs have replaced DSVs for most commercial work tasks, and the committee anticipates that ROVs will provide increasing support for marine science projects. However, there will continue to be certain vital exploration tasks that can be performed only by humans in situ.

Most contemporary DSVs require the on-site presence of a mother ship to provide logistical support for the vehicle and its personnel. The submersible's crew usually consists of a pilot and one or more observers. The observers are usually scientists, researchers, or technologists with an active part to play in conducting the mission. Due to limitations on human endurance and onboard power, mission times rarely exceed 8 hours, although some have extended to 16 hours or more. Emergency life support systems must be capable of operating for 72 hours beyond the maximum mission time.

The DSV places the crew directly at the site of interest. Visual observations are augmented for close-up inspections (less than 0.5-meter range) by video cameras, an important tool for direct observation (Robison et al., 1992). Most DSVs are significantly larger than ROVs used in comparable missions. Its size makes the DSV a stable platform to support viewing and manipulative tasks, including biological and geological sampling. However, because it is essential for the DSV to be large enough to accommodate several persons, it is more difficult to handle at sea and more difficult to position when performing tasks in restricted work areas. Essentially, DSVs are vertical probes with limited horizontal range; therefore, they are less suitable for large-scale

TABLE 2-1 Comparative Undersea Vehicle Capabilities

	DSVs	ROVs	AUVs
DEFINITION	Untethered, human-occupied, free-swimming, undersea vehicle	Tethered, self-propelled vehicle with direct real-time control	Untethered undersea vehicle, may be totally pre-programmed and equipped with decision aids to operate autonomously; or operation may be monitored and revised by control instructions transmitted by a data link.
DEPTH	Many to 1,000 m Few to 3,000 m Very few to 6,000 m One to 6,500 m	Very many to 500 m Many to 2,000 m Few to 3,000 m Few to 6,000 m One to 11,000 m	Several to 1,000 m Few to 3,000 m Very few to 6,000 m
ENDURANCE			
Time	Normally 8 hours, 24 to 72 hours max	Indefinite, depending on reliability and operator endurance	6 to 48 hours of propulsion May sit on bottom for extended periods
Range	< 50 km	Limited in distance from host ship by tether	350 km demonstrated; near-term potential 1,500 km, depending on energy source
PAYLOAD	1 to 3 people, 45 to 450 kg (100 to 1,000 lbs); adaptable to tools and sensors	45 to 1,590 kg (100 to 2,000 lb); adaptable to tools and sensors	11 to 45 kg (25 to 100 lb); adaptable to measuring equipment, tools, and sensors
SUPPORT			
Ship	Most DSVs require large ship support; ship size varies with DSV size	Depends on ROV size and mission requirements	Medium—depends on AUV size and mission requirements
Handling Systems	Depend on DSV size	Depends on ROV size	Similar to ROVs, depending on AUV size
Navigation Systems	Relative to seafloor or surface vessel	Relative to surface/seafloor	Seafloor and inertial navigation
STRENGTHS	Direct human observation and manipulation Real-time feedback to controller	Real-time feedback to operator, long endurance capability, low cost per operating hour	Potential for automated operations, ability to operate with or without human command and without tether; minimum surface support
LIMITATIONS	Large size, weight, and cost due to manned requirements Limited mission time Potential personal hazards	Tether cable potentially limits maneuverability and range	Energy supply Bandwidth of data link Capacity of internal recorders Limited work function complexity

mapping and surveying operations. As with other systems, advances in propulsion, energy storage, and manipulators will contribute to DSV utility and may even reduce their cost. Developments incorporating innovative uses of materials have already reduced the size and weight of the next generation of DSVs, and a system designed to bring a single pilot to the deepest part of the ocean is in progress (Hawkes and Ballou, 1990; Broad, 1993).

REMOTELY OPERATED VEHICLES

ROVs are by far the most common type of undersea vehicle; more than 1,000 ROVs have been built since their introduction in the 1960s. ROVs connect to a surface vessel or platform by a tether that carries power and control signals and feedback data from the vehicle. Originally developed for the military, ROV technology was further developed by the civil sector in the early 1970s, when private firms developed ROVs in response to the needs of the offshore oil industry. ROVs were one factor that enabled the offshore industry to move beyond diver-depth range. The results were reliable platforms serving a broad commercial market, with some technology transfer back to the military (McFarlane, 1987). ROVs continue to be used reliably in the offshore industry, and innovations in operational techniques and tool packages are expanding the scope of tasks these vehicles can perform (Langrock et al., 1992; Sucato, 1993). Nevertheless, ROVs are vertically operating systems that require significant surface support with attendant costs.

ROV manufacturing has been a highly competitive business. After attempts by several companies to compete in the rapidly growing offshore market of the 1970s, only one company, Perry Tritech, Inc., remains in the United States that builds full-size work platforms for the offshore industry. Worldwide, there probably are no more than five companies that have built more than one large ROV system. In the area of the smaller, low-cost ROV systems, a U.S. company, Deep Ocean Engineering, is the largest supplier among eight companies in the world that are in serial production of these vehicles.

Much of the commercial success gained by ROVs is due to the activities of service companies that operate vehicles under contract. Many of these companies began as commercial diving services, then gradually introduced ROVs as a lower-cost alternative to many underwater tasks previously performed by divers. Other companies began as ROV service organizations only, following the pattern set by former successful commercial diving contractors. In the past two decades, the primary driver for ROV technology advancement has been commercial-sector demand.

Light and medium-weight ROVs tend to have electrically powered thrusters. Heavy work-class ROV systems carry much larger payloads and tools, weigh up to 3,000 kilograms, and are fitted with hydraulic thrusters. These large ROVs found a niche in the offshore oil and gas industries and in the communication industry for underwater manipulation, cable burial, and inspection tasks. Large work-class ROVs are usually fitted with hydraulically powered manipulators, and some have protective cages that are used for launch and recovery. Nonmarine applications of ROVs, principally in nuclear and hydroelectric power plants and municipal water works, have evolved over the last two decades to provide small platforms for inspection of the radioactive, or potentially radioactive, components of nuclear power plants. The technical development and operational experience derived from the nuclear application has benefited the evolution and application of small-vehicle designs for undersea application.

The mobility of ROVs is often restricted by tether drag, and their stability can be affected by wave action on the surface vessel, which is transferred down the tether. Despite the constraints to horizontal operations within the sea imposed by the tether, ROVs have provided a platform for conducting in situ observations within the water column with little disturbance of surrounding sea life. ROVs have also extended the horizontal search range for larger, more limited survey vehicles such as the DSV *Alvin*.

AUTONOMOUS UNDERWATER VEHICLES

Although AUV research has been under way for several decades, the technological challenges and applications are such that these vehicles have developed more slowly than ROVs. A few systems were in operation as testbeds in the 1970s, but it was not until the 1980s, with the advent of microprocessors and associated software architectures, that these systems began to approach truly autonomous operations (Walsh, 1994; Michel and Le Roux, 1981). AUVs have potential advantages over other vehicles types. Because they lack tethers and carry no human occupants, AUVs permit sensing in areas where humans cannot go, such as under ice, in militarily denied areas, or in missions to retrieve hazardous objects.

For the past two decades, more than 75 percent of AUV development has been funded by the military, so continued development of this technology may be vulnerable to reductions in defense research budgets (Walsh, 1994). Further, because most of this work has been experimental or directed at military objectives, experience with AUVs in support of scientific missions has been limited, and there has been virtually no experience in the commercial sector.[1]

Several different types of AUVs have already been developed, each designed to respond to one of a variety of missions. The *AUSS*, shown in Box 2-1, addresses large area search and detailed inspection requirements. The lightweight AUV sensor platform (e.g., *Odyssey*; see Box 2-2) aims at fulfilling needs for medium-area surveys under ice and in deep water. The long-duration, deep-sea survey vehicles, for example, *ABE* (see Box 2-3), perform detailed inspections in deep water over long time periods (Michel et al., 1987; Walton, 1991; Walton et al., 1993; Bellingham et al., 1992; Bradley and Yoerger, 1993).

A recent emphasis in AUV development is on very small vehicles that can be deployed in large numbers to perform localized, relatively simple tasks such as shallow-water mine clearance. These vehicles could be very simple and inexpensive, using basic sensor suites and commercial off-the-shelf electronics. Working together, a constellation of such devices could accomplish the same work as a single, larger AUV. Because multiple units could be programmed to carry out the same task, redundancy would provide reliability. Although this concept is quite new and unproven, it has shown sufficient progress and promise to warrant continuing development.

As AUVs enter nonmilitary applications, it is likely that marine science will become a primary early mission and technology driver. As the technology matures, a number of applications in the offshore oil and gas industry, such as cable laying and inspection, and in a variety of other fields can also be envisioned (Fricke, 1992; Collins, 1993; Asakawa et al., 1993; Walsh, 1994). At present, virtually all AUVs are experimental prototypes or proof-of-concept vehicles. None is in routine operational service.

AUVs in existence today are vehicles with limited decision-making capabilities and endurance. The missions for AUVs call for simple data gathering, conducting searches, performing surveys, and laying fiberoptic cable. However, until more advanced capabilities evolve, missions requiring probabilistic decision making and true autonomy will be developed only for high-value objectives. AUVs are still in their infancy, and the lack of operational experience with these vehicles in the open ocean marks them as an immature technology with very important future potential. Even now,

---

[1] Examples of scientific work by AUVs (other than under ice) include the *Spurv* vehicle, operated by the University of Washington in the 1970s, which collects oceanographic data; the French *Epaulard*, which performed more than 300 dives between 1970 and 1990 (many to depths of 6,000 meters); the AUVs of the Russian Institute for Marine Problems in Vladivostok, which have conducted surveys at depths to 6,000 meters; and the *Odyssey*, which made dives in the Antarctic and in the Haro Strait off the state of Washington.

BOX 2-1
AUV Example: *AUSS*

AUSS, the Advanced Unmanned Search System locates and inspects objects on the ocean bottom over wide areas. Mission logic allows the vehicle to perform a side-scan sonar search, break off the search when it identifies a target for close-in optical inspection, and then resume the search where it left off after imaging the target. *AUSS* acoustically transmits its imagery and data to a vehicle supervisor on a surface ship. The vehicle uses Doppler sonar and a gyrocompass to navigate. It employs a cylindrical graphite epoxy pressure hull with titanium ends. It has demonstrated sustained searches at a rate of 1 sq-nm/hr, including evaluating individual targets discovered along the search path.

Owner:	U.S. Navy
Designer / Builder:	Naval Command, Control, and Ocean Surveillance Center
Operator:	Oceaneering Technologies for U.S. Navy Supervisor of Salvage
Purpose:	Deep ocean search
Depth:	6,000 meters (20,000 ft)
Size:	5.2 m long (17 ft) x 0.8 m (31 in.) diameter
Displacement:	1,230 kg (27,000 lb)
Speed:	6 kt maximum
Endurance:	10 hours @ 6 kt
Maneuvering:	Two vertical thrusters, two canted longitudinal thrusters
Energy:	20 kWh silver-zinc batteries
Payload:	Side-scan sonar, forward-looking sonar, electronic still and 35-mm cameras

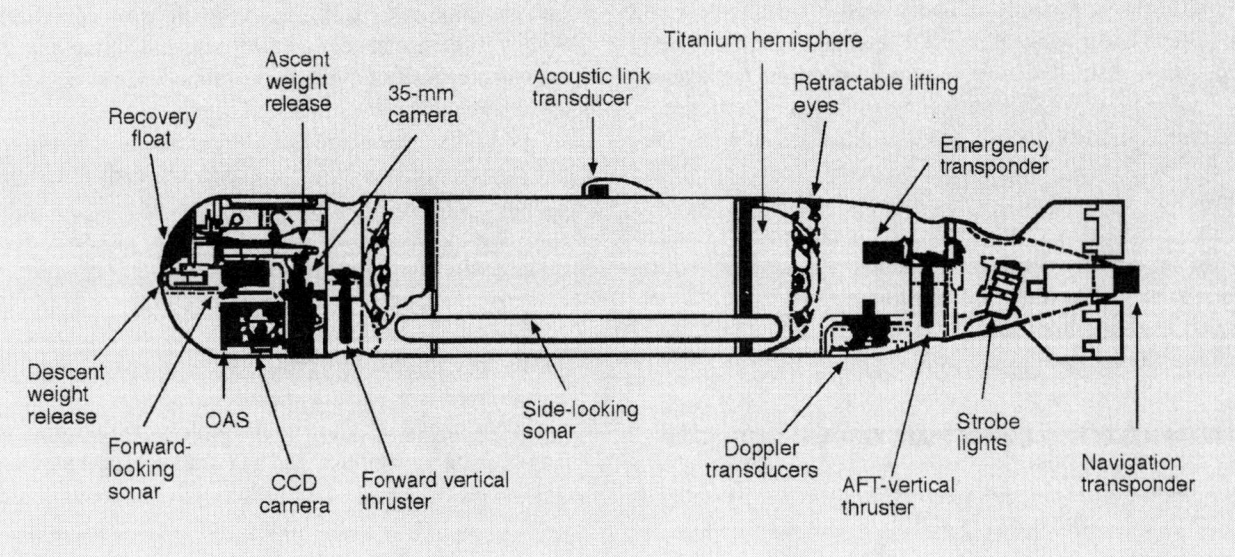

AUVs are capable of performing a number of clearly defined missions, but they have not been used because of their high initial development cost, lack of awareness of present vehicle capabilities, or lack of confidence by the potential user community.

Vehicle success in missions attainable with present system technical capability should provide both the experience and the support required for critical technology advancements to enable undertaking more complex missions. There is already evidence of this progress, as demonstrated by the AUV under-ice operations in the arctic (Bellingham et al., 1993) and in recent *Odyssey II* surveys over the Juan de Fuca Ridge (see Chapter 3). For example, high-level (low-bandwidth) human control of tasks requiring use of manipulators in real time is one important area under development. The human operator can command directly what is to be done with the object of interest within the water column or on the sea bed, and the vehicle-manipulator system will plan and carry out that command in near real time (Wang et al., 1992; 1993). This technology is similar to the approach being developed to deal with the time delays associated with robotic manipulation in space, and there has been some crossover of

BOX 2-2
AUV Example: *ODYSSEY*

The Massachusetts Institute of Technology Sea Grant AUV Laboratory has built six *Odyssey* class vehicles, of which five are presently operational. The five operational vehicles, designated type *Odyssey II*, are being used for Autonomous Ocean Sampling Network (AOSN) research and for development of a rapid event-response capability. A variety of field operations have employed the vehicles, including operations of the original *Odyssey* from the *Nathaniel B. Palmer* during an Antarctic cruise in January 1993. In March of 1994 *Odyssey II* was operated from an ice-camp in the Beaufort Sea. Deep-water operations of *Odyssey II* in summer 1995 culminated in dives to 1,400 meters deep and surveys lasting more than three hours. Vehicle development has been funded by the MIT Sea Grant College Program, the Office of Naval Research, the MIT, U.S. Navy Program Management Office 403, the National Undersea Research Program, and the National Science Foundation. The construction of the latest five vehicles has been supported by the Office of Naval Research.

Owner:	MIT Sea Grant College Program, AUV Laboratory
Designer/Builder:	Same
Operator:	Same
Purpose:	Science survey/AOSN research
Depth:	6,000 meters (20,000 ft)
Size:	2.2 m long x 0.57 m diameter
Dry weight:	115 kg present vehicles (maximum 160 kg)
Speed:	3 kt cruising, >4 kt maximum
Endurance:	6 hours at 3 kt (24 hours with maximum battery configuration)
Maneuvering:	One electric thruster aft of 4 control surfaces
Energy:	1.1 kWh Ag-Zn batteries (>5 kWh maximum)
Payload:	Sensors used include: video camera, conductivity, temperature, depth (CTD) instrument, optical backscatter instrument (OBS), mechanically scanned sonar, altimeter sonar, side-scan sonar, and acoustic Doppler current profiler.
Navigation:	On-board, long-baseline acoustic navigation, dead-reckoning, ultrashort baseline navigation also used for homing in Arctic missions.
Special Features:	Small and self-contained computer system ensure minimal support requirements. Free-flooded fairing provides large wet volume for addition of oceanographic sensors. Low cost through use of commercially available parts and selected manufacturing technologies.

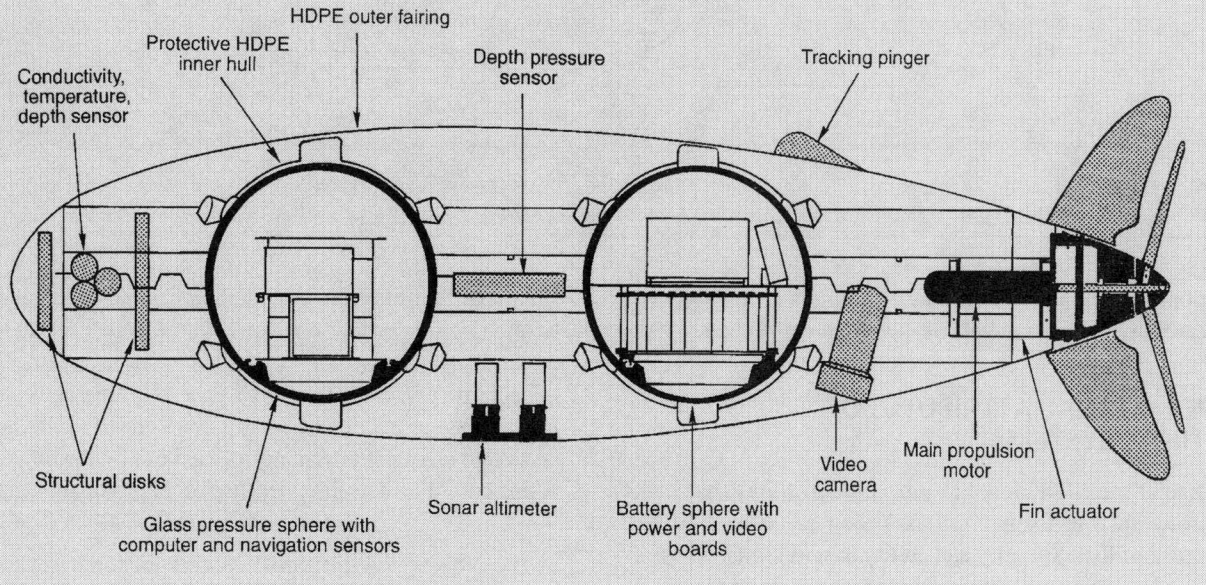

> **BOX 2-3**
> **AUV Example: *ABE***
>
> Woods Hole Oceanographic Institution built *ABE*, the autonomous benthic explorer for deep, near-bottom seafloor surveys. In 1995, *ABE* completed a geophysical survey on the Juan de Fuca Ridge at 2,200-meter depth, including magnetometer, conductivity, temperature, depth (CTD) instrument, and video survey. *ABE* has the ability to dock to a mooring and remain in "sleep" mode to perform preprogrammed, repeatable seafloor measurements over extended periods.
>
> | | |
> |---|---|
> | Owner: | Woods Hole Oceanographic Institution |
> | Designer/Builder: | Same |
> | Operator: | Same |
> | Purpose: | AUV deep ocean bottom surveys |
> | Depth: | 6,000 meters (20,000 ft) |
> | Size: | 2.2 m long |
> | Displacement: | 550 kg |
> | Speed: | 1 m/s |
> | Endurance: | 8–120 miles, depending on battery type and terrain |
> | Maneuvering: | Full hover, terrain-following |
> | Energy: | Lead-acid, alkaline, or lithium batteries |
> | Payload: | Stereo video snapshot TV, CTD, magnetometer, altimeter |
> | Special Features: | Low-power sleep mode, docking, closed-loop positioning, terrain-following, high- and low-frequency acoustic navigation. |
>
>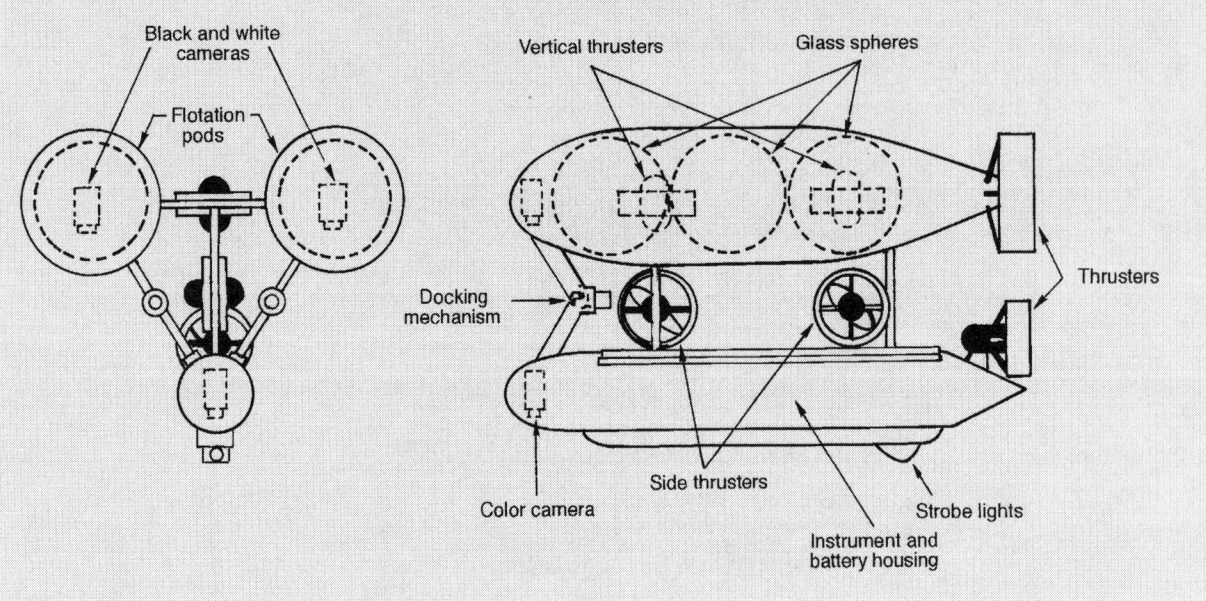

National Aeronautics and Space Administration (NASA) research with undersea vehicle development (Stoker, 1994).

OPERATIONAL ATTRIBUTES OF VEHICLE SYSTEMS

Each type of vehicle has inherent attributes that make it more suitable for certain tasks than are other systems. Because of their stability and ability to provide direct viewing for occupants/operators, DSVs are very good for observation and most work tasks requiring manipulators and samplers. However, because of their limitations of range and time on bottom or in the water column, which is dictated by the on-board energy source, human endurance, and the cost of support ship and crew, DSVs generally are not well suited for large-area search and survey or extended observation—tasks that are most efficiently carried out by towed vehicles at present.[2] Furthermore, because of their human occupants,

---

[2] DSVs have been used in some cases. An example was the *Challenger* search when the *Johnson-Sea Link I* and *II* had a large-area survey role in water depths of less than 1,000 meters.

TABLE 2-2 Current Undersea Vehicle Capabilities

Vehicle Functions	Requisite Characteristics and Capabilities	DSVs	ROVs	AUVs[a]
Reconnaissance	Forward observation, search, measurements	Good	Good	Good
Local survey	Small area observation and measurement, precise navigation	Good	Good	*
Broad area survey	Large area (up to 300 sq km) observation and measurement, medium geodetic or relative navigation accuracy	Limited	Limited	*
Waste site monitoring	Specific site observation, sensing, and water sampling	Good	Good	*
Mapping	Terrain feature survey, tied to accurate geodetic navigation, larger areas	Poor	Limited	Good
Search	Relatively large area coverage, acoustic and optical sensing, object identification, good navigation	Good	Limited	*
Inspection	Close-up observation, optical and other sensors, good vehicle positioning and stability	Good	Good	*
Observation	Similar to inspection, but implies real-time witnessing of dynamic events	Good	Good	Poor
Work	General tasks involving vision, object manipulation, and use of tools	Good	Good	*
Sediment sampling	Specific work task involving collection of material, including coring	Good	Limited	Limited
Installation/Retrieval	Placement or recovery of objects and instruments in/from specific locations	Good	Good	Good* Limited
Accident investigation	Observation, local area search, collection of material evidence	Good	Good	Limited
Waste disposal	Transport and placement of toxic materials in predetermined locations. May be large quantities or a deep site.	Poor	Good	Poor
Water quality measurements	In situ sampling and analysis in varying depths and locations	Good	Good	*

[a]Asterisks indicate that, while current AUVs are not suited to these tasks, developments are under way that could improve system capabilities to the point that the vehicle system should be suitable for the application.

DSVs are not suitable for operating in dangerous areas, such as in tunnels or around explosives.

ROVs, which are powered from the surface, have no real energy limitations. They are also generally stable. Viewing facilities for the human operator are good and continue to be improved, including stereo and new "augmented reality" compatibility.[3] Hence, ROVs are inherently suited for working for extended periods, performing local surveys, operating in high-risk areas, and passing large quantities of real-time sensor information back to a surface support vessel. However, due to tether drag, ROVs are limited in how far and how fast they can travel from their support craft, and they are less suitable for large-area, long-range search or survey and long, under-ice transits.

AUVs can move rapidly and, subject to limitations of the on-board energy storage, can generally traverse great distances relative to the other two types of vehicles. This makes them well suited for transporting sensors over large areas for surveys of various kinds. Some AUV systems have used a fiber-optic communications link for all or part of their operation. However, this is not a routine mode of operation, and, without a tether, the communication mechanisms for real-time human intervention are limited. Nevertheless, the narrow beam acoustic links have passed at least 50,000 bits per second (bps), and even low bandwidth links can pass some useful data. To extend the range of AUV applications, increasingly effective autonomous work systems are evolving rapidly along with the general field of robotics. Another present AUV limitation is that data transfer must wait until the AUV vehicle is recovered on board a mother ship. In response to this problem, acoustic telemetry schemes now emerging offer a hybrid arrangement where the human can intervene on a limited, non-real-time basis. New advances in task-level-control architecture will enable the operator to command tasks at a high level in real time (within the new bandwidth). AUVs are limited by the current lack of maturity of task-management architecture that can be placed onboard. Overcoming this limitation is the focus of present research at the Monterey Bay Aquarium Research Institute (Wang et al., 1993; Marks et al., 1994a; Wang et al., 1995).

Table 2-2 summarizes the foregoing discussion and includes a list of generic tasks that may be performed by vehicle systems. The committee has characterized the relative abilities of the different vehicle types to carry out these tasks; several qualitative descriptors are used, each representing the collective opinions of the committee and each based on considerations such as those discussed above.

EVALUATION OF THE STATE OF TECHNOLOGY

This section evaluates the state of the art and state of practice for vehicle technologies and assesses the potential for future developments. In the total-system context described

---

[3]For "augmented reality," the human can move an icon (i.e., a finger) inside the video scene to identify features (e.g., "that edge of that rock"), which the computer then uses in its task planning.

earlier, vehicle technologies are discussed here in relation to the various subsystems that are common to all subsea vehicles. The technologies described are typically applicable to several if not all types of vehicles covered in this report. This section groups the subsystems into two categories: those that directly support vehicle operations, that is, energy, propulsion, and control; and those that are related to payloads to support various missions, that is, sensing, survey, and manipulation. These two categories may overlap in some cases, but the distinction is useful for analysis. New developments with near-term usefulness are cited for each area, and the status of synergistic technology developments from other industries is discussed.

Vehicle Subsystems

Each subsystem and its driving technologies play a role in overall vehicle performance and contribute to the vehicle's capability to accomplish specific mission objectives. Lack of development in certain technology areas inhibits progress and further applications because they determine or facilitate vehicle capabilities. The technologies in other subsystems are highly developed, and further advancement will not appreciably improve the overall performance of the system. Accordingly, during the committee's evaluation, each subsystem was given an importance rating of "critical," "incremental," or "mature," depending on our evaluation of its impact on further vehicle development. These ratings are characterized as follows:

- Critical—Improvement in the subsystem will enable or create important new vehicle capabilities.
- Incremental—Vehicle progress can benefit from development of subsystems technologies in an evolutionary manner.
- Mature—Development has been successful and further improvement may occur, but development will contribute only marginally to improved vehicle performance, and improvements will be used only if they are cost-effective compared to current techniques.

Energy

Existing energy sources pose limitations for systems without cable connections (i.e., DSVs and AUVs), affecting system size, payload, and endurance. Energy limitations on AUVs are critical, and they are becoming more critical for DSVs because of the growing power demands of sensors, lights, computers, and manipulators. High-energy density batteries could lengthen missions and generally improve performance.

Energy sources are rated in terms of both energy and power. The most frequently used ratings are "specific energy" (watt-hours per kilogram) or "energy density" (watt-hours per liter). Batteries are usually used in underwater vehicles; numerous other energy technologies are also available, but they are more costly. The performance characteristics of available energy sources are compared in Table 2-3. The table is divided into four types of energy systems: secondary batteries, primary batteries, fuel cells, and heat engines. Secondary batteries are electrically rechargeable, while primary batteries are used for only a single cycle. (Ag-Zn may be included in either category; in its primary configuration it may be recharged as many as five cycles, which hardly counts as rechargeable.) Fuel cells are electrochemical devices that passively (without heat) react a fuel and an oxidizer to produce electricity; power levels are controlled by the amount of reactant injected into the cell. Many fuel cells are rechargeable, but not in the same way as secondary batteries; instead, reactant tanks are filled, and in some cases sacrificial metallic elements are replaced. Heat engines are generally closed-cycle, air-dependent systems that react fuel and oxidant in a mechanical cycle to drive an engine, which in turn directly drives the propulsion system or a generator to support electronic equipment. Table 2-3 is not intended to be all-inclusive. Instead it provides an overview of available energy technologies that can be considered for undersea vehicles.

Battery and fuel cell technologies developed for applications in space, automobile, and communications industries have not been adapted for use in undersea vehicles because of their cost, safety, immaturity in development, or incompatibility with marine missions. Cost is a primary factor and, as shown in Table 2-3, varies by orders of magnitude for different systems. Cost roughly increases as the energy density increases. For commercial AUVs and DSVs, the most advanced, high-energy systems are presently out of economical reach. These types of energy sources are found mostly in military systems where mission and endurance are primary factors that outweigh cost.

Factors important in selecting a battery include power density (the ability to deliver stored energy at the rate needed), outgassing properties, failure modes, reliability, ease and speed of recharge, and ability to operate over broad temperature and pressure ranges. Considerations of safety in handling energy sources, both aboard ship and aboard the vehicle, are critical and have limited the use of some chemistries, such as lithium, despite their high energies. (Some systems give off explosive gases during operation, and others may start fires if they fail.)

Current battery types used in undersea vehicles include standard lead-acid and nickel-cadmium batteries as well as silver-zinc and lithium-thionyl-chloride batteries. Figure 2-2 provides a helpful way to visualize the power and energy capabilities of various battery chemistries. For military applications, silver-zinc has been the *de facto* standard for over 20 years. Recently lithium-thionyl-chloride (see primary lithium in Figure 2-2) has seen some use because of its significantly higher energy rating. However, both silver-zinc and lithium-thionyl-chloride batteries are quite expensive ($300 to $1,000/kWhr) and have high life-cycle costs. Silver-zinc batteries are typically usable for 20 to 30 cycles,

TABLE 2-3 Performance Characteristics of Available Energy Sources

Technology	Specific Energy Wh/kg	Energy Density Wh/Liter	Cycle Life	Cost $/kWh	Maturity for Undersea Vehicles	Safety Concerns
SECONDARY BATTERIES[a]						
Lead Acid (Pb/Pb0)	35	90	800	50	Proven	H generation
Nickel Cadmium (NiCd)	55	130	1,000	1,500	Proven	Cd toxicity
Nickel Hydride (NiH$_2$)	60	150	10,000	2,000	Proven	High pressure H
Nickerl Metal Hydride (NiMH)	70	175	300	50	Proven	High pressure venting
Silver Zinc (Ag-Zn)	140	380	20	1,000	Proven	H generation
Silver Iron (Ag-Fe)	150	200	200+	500–800	Demo	H generation
Li-Solid Polymer Electrolyte (Li-SPE)	150	360	200	100–1,000	Lab	Lithium fire
Lithium Ion Solid State (Li-Ion-SPE)	150	360	1,000	100–1,000	Lab	None
Lithium Ion (Li-Ion)	200	200	2,000	500–1,000	Proven	Venting
Lithium Cobalt Dioxide (LiCoO$_2$)	220	300	50	1,000	Lab	Pressure venting, Li fire
PRIMARY BATTERIES[a]						
Lithium Sulfur Oxide (LiSO$_2$)	140	500	1	400	Demo	Li fire
Silver Zinc (Ag-Zn)	220	400	5	3,000	Demo	H generation
Lithium Manganese Dioxide (LiMnO$_2$)	400	450	1	200	Proven	Li fire
Aluminum-Seawater	450	400	1	100	Demo	N/A
Lithium Thionyl Chloride (LiSoCl$_2$)	480	500	1	300	Demo	Thermal runaway
Lithium Carbon Monofluoride (Li(CF)$_x$)	800	1,200	1	1,700	Proven	Li fire
FUEL CELLS						
Alkaline	100	90	400	5,000	Demo	Gaseous H and O fires
Proton Exchange Membrane (PEM/GOX/GH)	225	200	50	10,000	Demo	Gas H and O fire
Proton Exchange Membrane (PEM/LOX/LH)	450	400	50	15,000	Lab	H and O fires
Proton Exchange Membrane (PEM/SOX/SH)	1,000	883	50	5,000	Lab	N/A
Aluminum-Water Semi-cell (Al/H$_2$O/LOX)	1,200	800	1	10,000	Demo	H and O fire
HEAT ENGINES (Closed Cycle-Air Independent Propulsion Systems)						
Internal Combustion Engine	75	170	2,000	50–100	Demo	Fuel fire
Diesel Engine	125	75	1,000	100–200	Demo	Fuel fire
Brayton-Lithium Sulfur Hexafluoride (LiSF$_6$)	400	700	1	15	Demo	Fuel fire
Stirling	200	250	2,000	50–100	Proven	Fuel fire

[a] Battery parameters are based upon single cells; non-battery performance parameters are system level.

and lithium-thionyl-chloride batteries are primary batteries (not rechargeable). The higher energy batteries hold the potential for violent release of energy under certain circumstances and are not generally used in commercial and scientific applications (Harma, 1988; Moore, 1988).

Although higher energy systems are critically important to high-endurance AUVs, cost and safety must be primary objectives as new sources are developed. Newer secondary lithium chemistries hold the promise of reasonable production cost, high numbers of recharge cycles (approaching the lifetimes of lead-acid automobile batteries), no outgassing, benign failure modes, and energy densities better than offered by silver-zinc batteries. Development of these batteries is largely being driven by laptop computer and portable electronics applications, but they are presently being adapted to larger-scale batteries for automotive use and should be available for military and commercial undersea vehicles within three to five years.[4]

Recharging techniques for secondary batteries are being developed for a variety of user applications in the telecommunications, automotive, and undersea vehicle industries. The intent is to reduce the recharge time, extend the ratio of operating to charging time, improve battery cycle life, and promote personnel safety. Charging techniques such as pulse charging can be used to manage the recharge process and decrease recharge time, to reduce heat generation, and to minimize cell degradation. For undersea vehicles, some new

---

[4] This information is based on current Lockheed Martin Corporation programs and plans and on independent research and development related to energy sources as described in internally published documents (Gentry, 1995).

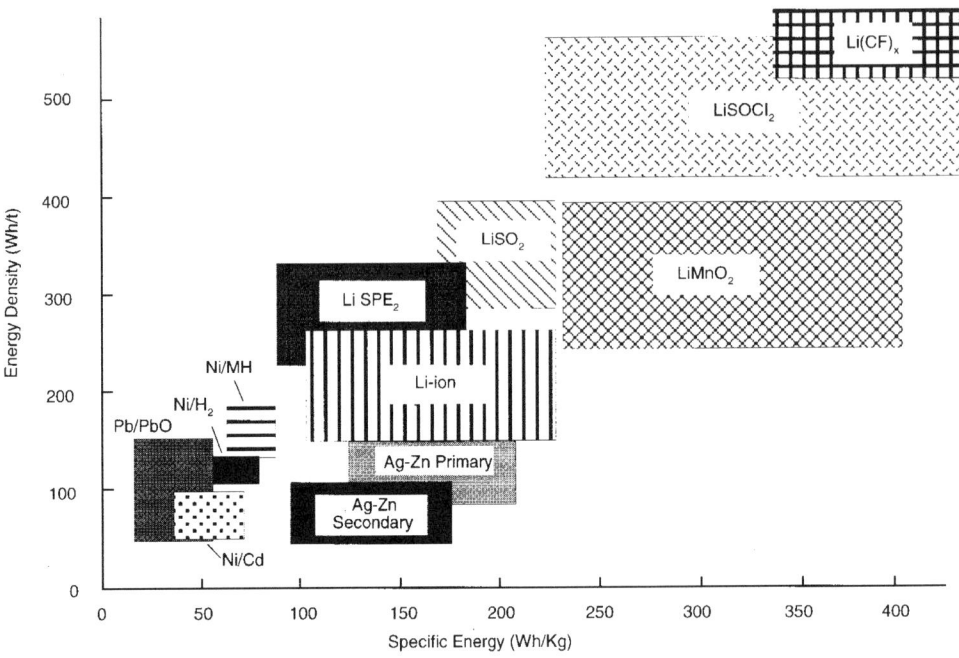

FIGURE 2-2 Battery cell comparisons. Source: Lockheed Martin Corporation.

lithium chemistries that emit little or no gas during charge and discharge are attractive because they can be charged in place (without recovery or vehicle disassembly). Other charging schemes include charging with solar cells, either by surfacing or by connecting to a subsurface charging station powered from the surface.

Other energy system developments include seawater batteries, which react metals with the oxygen dissolved in seawater. These batteries have been difficult to use because they produce very low voltage and power due to the limited quantity of dissolved oxygen in seawater. However, efficient dc-to-dc conversion can overcome some of these limitations. Since seawater batteries depend on dissolved oxygen, these batteries also may not be suitable in some parts of the ocean, such as in hypoxic areas (Blase and Bis, 1990). Military research and development efforts have explored using seawater batteries, but little work is ongoing in this area in the United States. The Norwegian Defense Research Establishment has built and successfully tested a long-range (approaching 1,800 km) AUV that employs magnesium seawater batteries with a specific energy of over 540 Wh/kg (Apel, 1993; Zorpette, 1994).

Metals can also be reacted with oxygen carried on board a vehicle. An aluminum-oxygen semi-fuel cell has been built and tested at sea with some success (Collins et al., 1993; Walsh, 1994). ARPA is developing a higher power-density version (Gibbons et al., 1991) of this type of cell that contains a chemically stable anode (aluminum), works at low temperature, and has environmentally benign byproducts, mainly Al_2O_3. The key concerns in this technology are oxygen storage and manufacturability. Canadian companies have also developed an aluminum semi-fuel cell, which uses pumped alkaline electrolyte and oxygen sources. This energy subsystem has been tested in an AUV (Stannard et al., 1995).

NASA has used alkaline fuel cells widely in spacecraft for over 20 years. One of these fuel cells was successfully demonstrated in a DSV in the late 1970s; however, cost and logistical problems limited further development for undersea use. ARPA is now evaluating a proton exchange membrane fuel cell (Meyer, 1993; Pappas et al., 1993). In this concept, stored hydrogen and oxygen are reacted in a fuel cell with the potential of rechargeability and specific energies of over 600 Wh/kg. The newer proton exchange membrane fuel cells offer many advantages over alkaline fuel cells, including lower cost, higher power capacities, improved tolerance to impurities in the reactant gases, and better long-term cycle performance. Energy capabilities of fuel cells are high, but they are limited by the difficulty of storing hydrogen and oxygen at high densities. To achieve truly high specific energies in fuel cells (> 1,000 Wh/kg) practical cryogenic (liquid oxygen) storage and solid hydride fuel are required. The logistics of reactant handling and storage continues to make cost reductions and practical usage of fuel cells elusive.

Isotope-based systems and small nuclear reactors have been proposed as ultrahigh energy sources (10,000 to 50,000 Wh/kg) for AUVs; however, these have not been implemented due to regulatory and cost restrictions.

Closed-cycle heat engines are another potential high energy source for AUVs and DSVs. These can use conventional or hydrogen fuels in combustion cycles (e.g., Diesel,

Brayton, Stirling) similar to engines developed for the transportation industry. A Stirling engine was tested successfully in an undersea vehicle by a Swedish company and is operational on board the French *Saga* vehicle.

The Office of Naval Research (ONR) has been pursuing improved thermal engines for torpedo propulsion for years. They have performed significant work in wick combustors, using liquid lithium reacted with sulphur hexaflorate to create a high temperature heat source for a Rankine or Stirling thermal engine (Hughes, 1995). In an attempt to achieve a more environmentally benign, refuelable, high-energy-density product (which is 80 percent of the energy generated by liquid lithium), the Applied Research Laboratory at Pennsylvania State University is investigating a wick combustor fueled with JP-5 (standard Navy jet fuel) and a lithium perchlorate oxygen source to drive the heat engine (Hughes, 1995). If successful, this approach may provide a less costly power source for AUVs in the future.

Most research and development in the field of energy storage occurs outside the undersea vehicles area. The committee anticipates that future energy system development applicable to undersea vehicles will derive mostly from the aerospace and automobile industries, where batteries and fuel cells are being evaluated for near-term use, and from the telecommunications and personal computer industries, where small-format lithium batteries are in development.

Energy systems are a low development priority for ROVs, whose performance is limited by other considerations. However, size, cost, and duration limitations related to DSVs and AUVs will be mitigated only when practical, safe, and readily available energy sources are developed. The advisability of making large investments in energy system research and development for commercial applications is questionable, and the decision must be made in the light of true development costs. Since most near-term AUV and DSV applications can be accomplished with existing and proven battery chemistries, research and development funds will be better spent in technology areas that are specifically marine, such as underwater navigation, acoustic communications, or subsea sensors. Energy systems will be advanced by industries such as the space, automotive, and telecommunication industries that have a more immediate need for them and can obtain development capital based on large markets for their products.

Propulsion

Most undersea vehicle thrusters now use fixed-pitch propellers driven by electric or hydraulic motors. The propeller configurations used derive from mature technology developed for ships and boats. However, optimization of propeller efficiencies continues as new undersea vehicle designs emerge. Brushless motors are often used for low cost ROVs and most AUV applications. Heavy duty ROVs at work in the offshore petroleum industry use hydraulic motors to power thrusters; power is supplied from an electrohydraulic unit mounted in the vehicle. The pumps, motors, and valves used in system integration are largely standard commercial products. Because of high propeller loading, DSV and ROV thrusters are frequently mounted in nozzles to improve efficiency. If nozzles are used on AUVs, they usually serve as propeller guards rather than as thrust enhancers.

Other thruster types, such as variable pitch propellers, cycloidal devices, and water jets have been abandoned because of complexity or inefficiency (Gangadharan and Krein, 1989). Oscillating foils, which function like a fish tail, are being studied and have achieved limited drag reduction (Triantafyllou et al., 1992). This concept may be applicable to AUV propulsion in the future.

Another interesting new direction in propulsion for long-distance observations is using controlled buoyancy or combinations of buoyancy and thrusters to propel undersea vehicles. The "Slocum" for example, is a concept that would use a heat engine, drawing on the ocean thermocline to adjust the buoyancy of an AUV; wings would provide lift and limited horizontal steering (Kunzig, 1996; Stommel, 1989). The concept offers the promise of low cost (tens of thousands of dollars, per vehicle, rather than millions); long range (2,000 km as a near-term goal); and increased endurance (50 days). The ultimate goal is to offer a fleet of low-cost, long-range AUVs that would operate simultaneously, taking oceanographic measurements with higher spatial and temporal resolution than are available with current techniques and at substantially lower costs. (The *Odyssey* vehicle, described in Chapter 1, is another low-cost vehicle that could be deployed in fleets, although it has a shorter range.) The Slocum buoyancy-adjustment mechanism (without lift or steering) has been tested at depths to 1,300 meters. The system currently being developed uses a battery-powered pump to enhance buoyancy control and propulsion (Webb, 1996).

Propulsion systems are a mature technology and a low priority for development since existing systems are adequate—although improvements in efficiency are always useful—for most underwater vehicle applications anticipated in the next decade.

Materials and Structures

Structural materials currently used for DSVs, ROVs, and AUVs have been adapted from the submarine and shipbuilding industries as well as from advanced aerospace programs. Design approaches derive from work by ship classification organizations, such as the American Bureau of Shipping, as well as from finite-element analysis techniques used in naval architecture and in many industries. Thus, with the exception of full-ocean-depth vehicles (11,000-meter depth capability), materials and design capabilities can be considered relatively mature. Improvements in materials tend to be more important for vehicles intended for deep applications because of the strength required to counter high pressures.

Improvements in material are also important for applications to AUVs, where lightweight materials can be translated into energy and payload for additional range and endurance and for work or sensing capabilities that require energy.

Currently, nonmetallic materials, including filament-wound epoxies, Kevlar, and graphite composites and ceramics, are used for military applications in both primary structures (pressure hulls) and secondary structures (fairings). However, the cost of some of these materials discourages commercial use. The Navy's *AUSS* uses graphite-reinforced plastic/epoxy for its main pressure vessel, but the future of such materials in pressure vessel applications remains unclear, primarily because of difficulties in manufacturing processes and high cost (Stachiw and Frame, 1988; Stachiw, 1993). Ceramic alumina cylinders have been tested for pressure housings and hulls, with potential weight reductions of 85 percent compared with titanium; however, economical manufacturing techniques are still under development (Stachiw, 1992; Kurkchubasche, 1992, DeRoos et al., 1993). Other ceramics being considered include silicon nitride, silicon carbide, and boron carbide materials (Ashley, 1993).

Advanced materials for fairings include graphite-epoxy layup or fiberglass constructions using a fiber-impregnated, high-density polyethylene that is also acoustically transparent (Sloan and Nguyen, 1992). *AUSS* uses this polyethylene material for its fairings. Advances in high quality acrylic and quartz glass will provide greater visibility to pilots of DSVs. Developments in important materials technology for vehicles aim to provide low-cost, lightweight, high-buoyancy materials for flotation. Sandwiched composite and syntactic buoyancy materials are being used to provide lightweight, high-displacement secondary structures. Although strength, density, and buoyancy are key design factors, longevity, corrosion resistance, and reliability also affect materials selection.

Design innovations have been demonstrated during development of new structures despite the relative maturity of conventional technology in this field. The two *Deep Flight* vehicle prototypes for a single-occupant, free-flying, full-ocean-depth DSV represent an innovation in alternatives for supporting human activity in the deep sea (Hawkes and Ballou, 1990; Ashley, 1993). *Deep Flight*'s pressure hulls are wound glass filament and epoxy matrix. Such a vehicle depends on advanced materials for structures to support its performance goals.

Russian and Ukrainian undersea vehicle programs have developed advanced techniques for fabricating structures of titanium, ceramic, and composite materials, according to two teams of experts who recently surveyed the undersea vehicle programs of western Europe and the former Soviet Union, under the auspices of the World Technology Evaluation Center (Mooney et al., 1996; Seymour et al., 1994).[5]

---

[5]These study teams included two members of this committee, J.B. Mooney and L.L. Gentry.

(Appendix B reviews the status of foreign undersea vehicle programs throughout the world.)

Another materials technology area of importance to system improvement is using coatings and other methods to resist biofouling or degradation of the vehicle's outer skin. Biofouling can create dynamic drag and interfere with the performance of skin-mounted sensors. This can be an especially critical problem for long-duration missions. Conventional coating systems that are used on surface ships may not be desirable for vehicles because the toxic compounds they use to kill organisms might cause chemical contamination of the vehicle's scientific sensors.

Navigation and Positioning

The success of most undersea vehicle applications depends on accurate navigation and positioning. Navigation is the function that continuously locates the vehicle within geodetic or relative coordinates and is critical to vehicle safety, operational productivity in real-time, and post-mission scientific and information processing. Positioning refers to the localized and more precise measurements often used to determine specific distances relative to some fixed point. For example, vehicle work in the offshore oil and gas industry frequently involves precision measuring and positioning of equipment relative to installations on the seafloor. When operating a vehicle in a localized area, most contemporary navigation and positioning systems make use of acoustic transponders such as the long-baseline networks widely used in many types of deep-water work. Systems of this type use bottom-placed transducers in array fields with typical transponder separations of up to 4 km and can offer accuracies of 1 meter at frequencies of 26 to 36 kHz. Recent developments in acoustic positioning include a high-frequency, high-accuracy system that determines the position of a vehicle with an accuracy of a few centimeters in a bottom-placed transponder field. Other systems, which utilize transducers mounted on the surface ship and a transponder on the vehicle (short baseline), do not require transponders on the seafloor. These systems are widely used for navigating vehicles relative to a support ship. Combinations of these acoustic systems are used to maximize the advantages of each for best navigational accuracy for specific environmental conditions.

Numerous other acoustic and nonacoustic sensor technologies are used on the vehicle to enhance navigation and positioning. Simple video cameras are useful, especially for ROVs and DSVs, when operating near the bottom of a structure and can provide the operator a reference for motion. Computerized image processing techniques have been developed that can use information from video cameras to navigate vehicles automatically (Wang et al., 1992; Marks et al., 1994a, 1994b). Further developments of this type will enhance the value of video as a navigation aide, especially for AUVs, where precise autonomous, near-field navigation is still under development. Other optical sensors such as laser imaging distance and ranging (LIDAR or laser radar) and line-scanning lasers are becoming useful for short-distance ranging and imaging and have recently been developed in low-power versions (< 200 watts) that will permit their use on AUVs and DSVs.

Scanning or multibeam sonars are used to provide operators with images of obstacles and terrain in the immediate area surrounding the vehicle. These systems are very popular and have developed to the point that high-frequency, high-resolution systems are reliable and economically available on the commercial market. Current developments in sonar and signal processing include obstacle-avoidance sonars that can construct a terrain map and guidance strategies for optimum pathfinding.

Navigation over long distances or for prolonged durations is generally a requirement for AUV missions and is critical to mission success. AUV navigation systems typically use magnetic or gyro compasses and a velocity sensor to provide dead reckoning.[6] Inertial navigation systems, aided by Doppler sonars, are an advanced implementation of this technique. These systems, derived from extensive use on aircraft and spacecraft, provide inertial navigation, which is then corrected by velocity estimators and by position fixes, as available. Estimated vehicle speed is obtained from current and flow sensors, Doppler sonars, or correlation sonars. Doppler sonar uses reflected echoes to provide highly accurate measures of motion relative to the bottom or fixed points in the water column. Similarly, correlation sonars accurately (+/– 0.1 kt) measure speed relative to the bottom (Grose, 1991). Doppler correlation sonars can "bottom lock," referencing the vehicle's motion to the bottom, from altitudes of 3,000 meters or more. The advent of small inertial devices, such as ring laser gyros (a solid-state version of the conventional rotating gyro), are making this type of navigation increasingly useful as accuracy goes up and cost goes down (Moore, 1991; Ezekiel, 1991). The result is that velocity-aided inertial navigation systems are now available that provide accuracies on the order of 0.1 percent of distance traveled, and further improvement will result from integration with better location and true movement sensors. As these improvements continue, costly, time-consuming transponder fields will become increasingly unnecessary.

The greatest advances in undersea navigation in the near future will come not from any one isolated type of system but from integration of an increasing number of systems and components. Most of the above techniques benefit greatly from position referencing to the Global Positioning System. Recent work in the combination of inertial units with ultra-short baseline transponder systems and Doppler sonars has shown that combining sensors with different characteristics can synergistically improve navigation performance (Hutchison and Skov, 1990; Hutchison, 1991). Accuracy on the order of 0.05 percent of distance traveled are achievable with inertial navigation systems and Doppler sonars, which are becoming available at moderate cost.

Guidance and Control

Navigation, guidance, and control functions are often separated for discussion, as is the case in this report. This modularization assists in understanding the complex operating concept for undersea vehicles and is also helpful to the vehicle designer. In practice, however, these functions are highly interactive and, in fact, use many common sensors and processors. Thus advances or improvements in one function normally are linked to advances in other functions. For example, the development of a highly accurate, long-duration navigation system would be useless without guidance and control capabilities that support mission intelligence and reliable navigational capabilities.

Guidance and control of an undersea vehicle are generally implemented in a layered or hierarchical architecture. Guidance involves higher-level mission management activities, such as planning and directing vehicle movement through the water column; control operates at a lower functional level to interact with specific equipment on the vehicle. The control level includes the closed loop functions (autopilot) that provide stable, controlled operation of the vehicle. The control level receives orders from guidance and, in turn, commands physical actuators, propulsors, and effectors to maneuver and operate the vehicle in a manner that accomplishes higher-level guidance objectives.

In the early days of undersea vehicle development, maneuvering depended almost exclusively on the direct manual control skills of human pilots, and all higher-level planning was accomplished by the pilot. With ROVs, pilots worked primarily from video images, using visual references to keep track of vehicle and tether location. Later, automatic heading and depth controls became common on most vehicles because of the evolution of reliable sensors, modern computing equipment, well-understood control algorithms, and efficient software. Tracking systems, imaging sonars and inertial navigation systems also improved the human operator's ability to determine vehicle position in geodetic or local reference coordinates, thus enhancing vehicle guidance and control.

Continuing improvements in navigation and control technology permit automation of all vehicle motions. A vehicle with full automation and control of movement and direction can hover for long periods and can follow preplanned track lines precisely while under "supervisory control," that is, with the human operator providing high-level, task-oriented commands rather than exercising direct control over all functions of the vehicle (Yoerger and Slotine, 1987; Wang et al., 1993). Vehicles equipped with such capabilities have

---

[6] Dead reckoning is defined as the finding of location using compass readings and other recorded data, such as speed and distance traveled, rather than astronomical observations.

performed detailed, three-dimensional scientific surveys of archaeological sites and deep ocean hydrothermal vent plumes, dam and nuclear reactor inspections, "hands-off" docking on oil field structures, and automated operation of valves (Somers and Geisel, 1992). Techniques are also being developed to allow vehicles to hold position based on video, laser, and acoustic imagery and to use imagery for guidance (Marks et al., 1994a, 1994b; Negahdaripour, 1993; Wang et al., 1992). Many advances in control depend on improved understanding of vehicle dynamics and improved signal processing algorithms for closing feedback loops around sensors to form servo loops (Yuh, 1990; Healy and Leonard, 1993; Fossen, 1994).

Future developments may extend these capabilities in several directions. Improved navigation that combines inertial and velocity measurements (as described in the previous section) currently being developed for military AUVs will enable precise automated vehicle motion without the need for a transponder network. Control systems combined with sensors that detect cables, pipelines, hydrocarbon leaks, or other pollutants allow highly efficient automated tracking and surveying (Greig et al., 1992). Likewise, improvements to in situ sensors for oceanographic parameters and chemical samples, combined with advanced vehicle control systems, will allow scientists to map distributions with unprecedented sampling density at reasonable cost, perhaps with multiple vehicles (Triantafyllou, 1992; Curtin et al., 1993). At the heart of these improvements is the ability to integrate navigation, guidance, and controls with sensors, using modern hierarchical architecture techniques to enhance accuracy, efficiency, ease of human task-level control, and reliability of vehicles for a wide range of mission needs.

A significant step toward achieving these integrated goals was demonstrated in a series of 13 dives over the Juan de Fuca Ridge by MIT's AUV *Odyssey II*. *Odyssey II* reached a depth of 1,400 meters and ran surveys designed to characterize spatial variability of temperature and salinity in three dimensions. Navigation was provided by a long-baseline acoustic navigation system. In a 3.25-hour dive, the vehicle excursions centered on a thermocline at 45 meters beneath the sea surface. To provide an understanding about temporal evolution in the survey volume, the grid survey was preceded and followed by vehicle paths crossing the survey volume.[7]

As indicated above, developments in guidance and control, at all levels, are critical for progress in AUV applications where robust mission management is key to reliable and repeatable performance. Artificial intelligence techniques are being applied to offer AUVs an interpretive logic capability based on processing probabilistic data.

In the future, AUVs should be capable of pursuing tasks that have abstract descriptions; for example, finding and following a chemical gradient or surveying a given area with the ability to replan and reconfigure the mission based on a wide range of changing internal and external factors. Included in these tasks are a number of lower-level operations, including obstacle avoidance, homing and docking, and following terrain, as well as manipulative tasks that involve control of the vehicle/manipulator system to carry out a command.

Failure detection and recovery are perhaps the most critical operations and the most difficult. The vehicle must be able to sense when one or more of its subsystems have failed and must be capable of reconfiguring its controls and replanning the mission in real time to work around the problems; accomplishing the highest priority objectives; and in the worst case, aborting the mission in the safest manner. The vehicle must also be able to handle high-level failures such as reattempting a docking operation that fails the first time (Ricks, 1989; Perrier and Bellingham, 1992). "Layered control" is one approach to this problem. Vehicle software provides commands for a set of quasi-independent "layered" behaviors, such as "detect collision," "hold heading," or "head to way-point." Layered control has demonstrated some success as an overall philosophy for AUV programming (Bellingham and Leonard, 1994). A Navy-sponsored project is developing an intelligent, fault-tolerant vehicle guidance and control system, and system testing and demonstration are planned. While much of this development is directed toward specific use by AUVs, it is supported by complementary work of the computer aerospace and automated manufacturing industries. Continuing advances in task-level control architectures and higher bandwidth communications have resulted in robots that respond directly to graphical task-level human input. These robots use an advanced form of "telerobotics," or control from a distance, which until recently allow only a "joy stick" human interface.

For many mid-water tasks the vehicle and its manipulators need to be controlled as a single moving system. The new capability called "object-based task-level control" enables the direct human command to the task that will be performed; the control system then plans and executes the task. Because of the much lower bandwidth required for task-level commands, object-based task-level central will enable near real-time control of AUVs, which will be a powerful new capability (Wang et al., 1993). Techniques for vehicle control are continuously being improved. Research and development activities are being directed toward adaptive systems that can successfully control a vehicle with widely varying characteristics (e.g., mass or hydrodynamic coefficients). The use of sliding-mode controllers is one approach; another involves intelligent systems that estimate vehicle characteristics in real time.

---

[7]This deployment was the latest of seven field operations of the *Odyssey* series of vehicles that included untethered operations in the Antarctic and the Arctic under ice. Four new *Odyssey* vehicles (the *Odyssey IIb* class), were used for docking system development and acoustic communications experiments in 1995 and will be used for ONR field programs (Bellingham, 1995).

Improved guidance and control techniques can enhance the capabilities of ROVs and DSVs and are crucial to the success of AUVs. Improvements in fault tolerance can permit vehicles to complete missions after sensors, actuators, or processors fail or degrade. Another important direction for AUVs is to move beyond simple way-point control to permit the vehicle to pursue tasks from more abstract descriptions.

Data Processing

Undersea vehicles typically require two types of data processing, payload and vehicle management. The payload processor collects, processes, compresses, and records the data produced by the vehicle and its sensors, often on disk in the vehicle itself or on a support vessel. The data are generally recorded during operations and processed afterward, especially in scientific applications. Data compression is essential when recording devices or the uplink bandwidth are limited and data volumes are large. The payload processor also can perform processing to augment and fuse the data that are collected; for example, the vehicle data can be matched to the image from a sonar, and the fused result gives an accurate picture of the situation encountered by the vehicle at a given time and place. The advent of fiber-optic communications and advanced sensors for ROVs has allowed transmission of large volumes of data up the tether for data logging, management, and display.

The second type of data processing is performed by the vehicle management computer, which can be located on board the vehicle or on a support vessel. These data may also be processed in real-time for use in decision making, such as for mine detection and target classification applications. The vehicle-management computer typically performs all the housekeeping functions necessary to keep the vehicle in motion along the prescribed path. The data can be used to control vehicle functions such as thrusters, control surfaces, valves, and manipulators in real time. As the human operator becomes more removed from the vehicle control loop, and as tasks become more automated, the performance of vehicle-management computers becomes critical to mission success.

Another key issue in the data processing chain is calibrating the sensors used to make measurements and detecting a failed or faulty sensor. The vehicle-management processor must continuously monitor the sensor output for validity and presence. Sensor redundancy can help considerably in this process; however, deciding when a sensor has failed and when to switch to a backup sensor is a difficult process without an operator in the loop. In scientific operations using tethered systems, the sensor outputs are continuously monitored by the operations team, and the team decides whether to continue the mission or retrieve the system and initiate repairs. This process must be automated in AUVs; therefore, it is a critical area for improvement. Intelligent systems are currently being developed that will be able to monitor and compare sensor outputs. Upon detection of a failed sensor, these systems will make an informed decision about the goals of the mission and cause the vehicle to surface or continue depending on criteria that have been preprogrammed by the scientist or operator.

Developments in data preparation, fusion, presentation, and analysis are necessary to fully use and understand data collected by undersea vehicles. Fusion of many different types of data, including sonar, video, still images, water column measurements, and vehicle positioning data, must be available for scientific evaluation (Rosenblum et al., 1993; Gritton and Baxter, 1993). In the past, post-mission processing of data was performed by a human, who correlated photos, sonograms, strip charts, and computer data in a time consuming and not always accurate attempt to evaluate the vehicle's mission. Current advancements in post-mission processing, which incorporate simulation and display of the data collected from a mission using three-dimensional graphic software tools, aid the scientist in accurately reconstructing the mission and evaluating the data. Real-time map and chart construction fuses vehicle positioning, terrain, and targets encountered during a mission and allows faster evaluation of post-mission data (Howland et al., 1993). These techniques also aid the pilot and the observer in constructing a mental image of the subsea environment while they are working. Designers of vehicles for scientific applications will increasingly build data management into their basic design philosophies to make the data collected by undersea vehicles more useful and more easily understood (Newman and Robison, 1993).

Some help for vehicle designers may be provided by software systems designed for data processing in military applications. Current studies on the correlation of satellite telemetry and image data can be directly applied to the undersea world in the future. Commercial systems are also emerging that can handle diverse vehicle data. The new PC processors are now powerful enough to display complex data in real time, complete with three-dimensional colorized representations of detailed images. Contemporary commercial software tools make plotting numerous pieces of data on multiple charts easier. New display techniques, including virtual presence, will further ease the task of understanding and interpreting data from the undersea environment by enabling scientists to position themselves aboard the vehicle and fly the mission, experiencing firsthand the situations surrounding the data as they are collected (Gwynne et al., 1992).

Enhancements in undersea vehicle capabilities are closely tied to advancements in microprocessors and computer science. In the past 10 years processing signals and managing underwater vehicle systems have progressed from implementing a single desk-sized minicomputer to incorporating many, in some cases hundreds, of printed circuit board processing elements. This improvement enabled major reductions in cost and size and has increased the availability of microcomputers that continuously improve in terms of low-power operation and computational throughput. The microprocessor boom, in conjunction with the shift toward distributed networks of specialized computers, has resulted in a revolution in processor system capabilities for undersea vehicles.

DSVs and ROVs have benefited from increased reliability of on-vehicle data processing systems and advanced operator displays that are adaptable to various missions and operators. Miniaturization of control systems has allowed integrated operations of DSVs and ROVs, extending their combined capabilities beyond that of a DSV operating alone. Increased capabilities for storing and managing data aboard DSVs has also enhanced data integrity and accessibility for scientific applications. Underwater vehicles have also borrowed from advanced computer science in the move toward distributed processing of vehicle sensor information through "smart" sensors that are directly integrated with a microprocessor. This has the additional benefit of standardizing sensor protocols and message formats and increasing design simplicity for human applications. Data incorporation from multiple sensors through a single analog to digital converter has given way to on-vehicle networks that extend through the data transmission system to the support platform to provide better displays of sensor information and increase reliability.

AUV missions are clearly the most computation-intensive of undersea vehicle applications; yet these are easily being implemented with current computational capabilities. Over the past decade, the computing revolution has resulted in order-of-magnitude increases in processor capacity every few years at continually reduced power levels. A central processor the size of a small loaf of bread, weighing under 4.5 kg and drawing less than 10 w, can implement all guidance, navigation, and control functions required by an AUV.[8] Future AUV missions will demand the higher levels of onboard signal processing and data processing that are associated with increased levels of autonomy. However, it is expected that developments in semiconductor materials, improved board geometries, and more efficient operating systems for compilers will easily meet the computing requirements. In addition, the use of advanced paradigms, including artificial intelligence, fuzzy logic, and neural computing, is becoming more mature, and these paradigms will implement efficient sensor-based perception and data fusion for object interaction and advanced fault detection, isolation, and management.

Current work in signal processing is benefiting acoustic communications and sensor signal processing. The incorporation of low-power, algorithm-specific processors with high processing rates provides the required throughputs for separating signal from noise and integrating signals into meaningful information. Advancements include distributed and parallel processors to interpret on-vehicle acoustic, video, and laser imagery when evaluating objects in the environment.

Vehicle design, analysis, simulation, and verification, especially for AUV applications, have benefited from advancements in computational capability. Vehicle development and construction have progressed from empirical hand-drawn designs to completely computer-developed and maintained representations that can facilitate automated manufacturing. Analysis is enhanced through solid modeling and simulation that approaches real time in complex tasks such as fluid flow and vortex analysis. At the far extreme of capability are the enhancements being made in ARPA's Simulation Based Design Program, which extend the current capability to allow three-dimensional virtual reality visualization of the vehicle and its subsystems, combined with physics-based modeling, to evaluate performance before physical construction.

As mainstream computing hardware and software evolve, these advances can be incorporated into underwater vehicles to make them more capable and reliable. In particular, techniques that combine diverse types of data will allow scientific and commercial data products to be produced more quickly, more cheaply, and with higher quality. In addition, automated techniques to monitor and manage sensors will be vital to permitting AUVs to produce high quality data sets without intervention.

Communications

Communication between human operators and the vehicle system—to receive control signals, report mission status, and transfer sensor data—encompasses several technologies, the use of which depends on the vehicle type. In general, AUVs use communication transmitted acoustically through water at 8.075 kHz and 27 kHz (operated half-duplex) as the most common frequencies. ROVs use an umbilical cable or link that contains coated, shielded, twisted-pair and/or fiber-optic members.

Communication capability with tethered vehicles (ROVs) has significantly improved with the transition from copper-based to fiber-optic-based systems. This advancement has closely followed developments in the telecommunication industry. The emergence of fiber-optic data transmission and communications technology has increased the capability of ROVs to pass enormous volumes of data through the tether for data logging, management, and display. As the volume rate and reliability of the data have increased, so has the data value to operators and users. It is now possible to tie sensor information regarding temperature and chemical composition, for example, to precise, specific locations relative to geological resources. The collection of sensor information and geolocation has also become critical in military mine countermeasures and explosive ordnance disposal applications.

---

[8]This reference is based on an internal Lockheed Martin Missile and Space Company report documenting the development and testing of the common guidance computer being used in a proprietary program.

ROVs generally require transmission rates greater than 1 megabyte for video and data channels, which is easily satisfied by twisted wire pairs in tethers of less than about 1,000 meters in length. Optical fibers, which are included in tethers for very long distances or for high transmission rates, exceed coaxial cable performance by several orders of magnitude, permitting the use of a large number of digital and video lines. At present, multimode fibers are used for umbilicals 1,000 meters to 3,000 meters in length, and single-mode fibers are used for longer umbilicals. Single-mode fiber-optics and associated electronic technologies are evolving rapidly, and it is likely that single-mode fiber-optics will become dominant in all umbilical lengths. In the fiber-optic communications system, data are passed along the umbilical by line driver receivers, laser diodes, and other off-the-shelf devices. Both digital and analog data streams are used, as are error-code checking strategies adapted from other applications.

Fiber-optics can be used for special vehicle applications, including providing a disposable telemetry link from the vehicle as it travels (Brininstool and Dombrowski, 1992; Grey, 1992). Distances in excess of 185 kilometers (100 nautical miles) are achievable without using repeaters. The difficulty of repair in the field has been a limiting factor in the proliferation of fiber-optic systems, but improvement in field maintenance is removing this constraint.

Except for special hybrid configurations, where fiber-optics are employed, autonomous tetherless vehicles cannot benefit from the advantages provided by fiber-optic communication links and must rely on the sound-transmission characteristics of the ocean. One of these characteristics, the speed of transmitting the acoustic signal, results in a lag between the transmission and receipt of the command data or signals, the resultant action, and the operator's perception of the effect of the action. This is a similar problem (but much greater in time scale) to the problem challenging teleoperations for the NASA Mars Rover Program. Long transmission delays are generally overcome by some level of autonomous behavior within the vehicle and supervisory control by the operator (for example, commands are given by the operator to pick up an object rather than transmitting detailed manipulator orders). This level of abstraction has come about only recently with the development of high-speed computers and microprocessors.

Although data flow via acoustic links is limited by the seawater medium to very low bandwidths, using present technology and practice acoustic telemetry can carry control data and transmit image data from untethered vehicles or between networks of seafloor instruments and vehicles. Recent developments in signal processing and communications can be applied to exchange data at a reduced rate. Until recently, acoustic communications with submersibles has been limited to voice, such as the use of the underwater telephone. The range of this voice or Morse Code communication path is inversely proportional to the transmission frequency and limited by the transmission power. Acoustic communications have proven successful for long, straight paths in deep water (Mackelburg, 1991). In shallow water and with horizontal signal paths, however, the acoustic qualities of the ocean, combined with reverberations caused by reflections off the surface, seafloor, and other obstacles make the problem much more complex (Catipovic, 1990).

Current techniques for acoustic communication can deliver rates approaching 20,000 bits per second (bps) in deep water and about half that in the more difficult environment of shallow water at horizontal ranges. This technology is changing rapidly, especially for systems that communicate with free-swimming vehicles. Pioneering attempts at acoustic digital data communications were developed by the Naval Command, Control and Ocean Surveillance Center, San Diego, California, for use in supervisory control of the AUSS vehicle. The vertical communication path enabled transmission of command signals at a rate of 2,400 bps and freeze-frame video images were transmitted using data compression at a rate of 4,800 bps. The incorporation of computationally capable, low power, algorithm-specific signal processing by Woods Hole Oceanographic Institution and Northeastern University (Stojanovic et al., 1993; Stojanovic et al., 1995) has permitted significant improvements in digital acoustic communications, approaching a limit of approximately 3 bps/Hz. Catipovic has successfully demonstrated long-distance, shallow-water communications at burst transmissions of more than 30,000 bps with a continuous throughput more than 8,000 bps out to approximately 3 km.[9] The realization of these data rates allows for real-time acoustic transmission of data at significant baud rates to enable real-time vehicle control limited by the speed of sound through the water. A series of acoustic modems can then be used to control a vehicle in arctic waters (Bellingham et al., 1994) or to send data to a central collection buoy from a distributed network of ocean-bottom sensors (Robison, 1994).

Acoustic telemetry has limited range by itself but can be combined with other types of telemetry to provide longer ranges. Other complementary types include satellite telemetry or cabled telemetry from seafloor acoustic installations. A vehicle could surface periodically to transmit data to low earth-orbiting satellites, although practical problems concerning vehicle stability, antenna design, and power consumption must be addressed. A vehicle could also communicate acoustically to a network of moorings, some of which are equipped with surface buoys and satellite links. Likewise, an AUV could communicate acoustically to a hard-wired installation such as a seafloor observatory or other hydrophone array.

---

[9]This demonstration was conducted during the American Preparedness Defense Association symposium, November 1, 1994, by the Woods Hole Oceanographic Institution in Narragensett Bay. The demonstration was supported by ONR and ARPA. An AUV was maneuvered while transmitting images acoustically to a buoy that transmitted the data via radio frequency to a conference hall, where the images were displayed (Catipovic, 1996).

Significantly higher data-rate transmissions can be achieved using pulsed lasers. An ARPA-sponsored project demonstrated 100 megabytes per second data transmission from an autonomous vehicle to a submerged submarine in 1992.[10] The attenuation of light in the ocean, however, limits the pulsed laser range to approximately 100 meters or less.

An important area of communications for undersea vehicles is the use of satellite or radio frequency links. Both the Woods Hole Oceanographic Institution and the Monterey Bay Aquarium Research Institute have used this method to display science results from ROVs working offshore in real time at the laboratories on shore. AUVs may also benefit from this technique by docking at subsea data-transfer stations periodically to recharge batteries and transfer data for transmission ashore via surface buoys.

This is a very active field, drawing on advances in the electronics and telecommunications industries. Significant advances are also being made in sonars and acoustic sensors and in understanding acoustic oceanography. The impact on DSVs and ROVs will be incremental. However, as improvements are made in communication systems, especially acoustic links, and as the new systems become cost effective, they will have a major impact on AUV performance capability.

Payload Subsystems

Payload systems are those carried by the vehicle to perform mission tasks. Payload requirements influence the design of the basic vehicle platform for specific missions. The challenge for the payload designer is to achieve the maximum system performance within size and energy-consumption limits. Payloads systems include:

- work systems (e.g., manipulators, end-effector tools, tool racks, and bins for storing samples)
- sensor systems (acoustic, optical chemical radiation, gravity, and magnetic field sensors; fluourometers and transmissometers; and conductivity, depth, temperature sensors)
- current meters
- payload power systems (often separate from the main power system, so that a failure of payload power will not shut down the full system)

The number, size, and weight of payload systems affect vehicle size, mass, and propulsive power requirements. The number of energy-consuming payload systems helps determine the total energy system requirements; energy efficiency is important.

Work Systems

Manipulator arms are used on both DSVs and ROVs to accomplish common scientific and industrial tasks;

---

[10]Tests were successful, but no subsequent test reports have been publicly released (Pappas, 1995).

manipulator development is likely to result in incremental increases in performance capability for these vehicles. Manipulator use on AUVs is still embryonic, but improvements could significantly enhance AUV capability for performing a greater breadth of operations. Current practice involves rate or master-slave manipulators, where the operator (located inside a DSV or on a surface vessel controlling an ROV) operates the arm by throwing switches or by moving a miniature version (the "master") of the manipulator on the vehicle (the "slave"). Typically, modern hydraulic arms on large ROVs can lift hundreds of kilograms, even when fully extended.

New control techniques drawn from space developments will allow the human operator to command directly at the task level what is to be done with the object of interest, and the vehicle-manipulator system will respond by carrying out that command. The operator needs no special "crane operator" skills, and a scientist or the field engineer can play the operator role. The operator can then focus completely, in real time, on the task itself and the objects to be manipulated, whether they are science samples, cores, or equipment (Wang et al., 1993).

Manipulators need a device, called an "end-effector," to perform the actual task. End-effectors are often general-purpose hands or grippers, but they can also be special-purpose power tools, such as drills, cutters, or wrenches, especially in offshore oil applications. These tools are often grouped into "tool packages" that allow the vehicle to use several different tools during one dive. Whereas formerly it was considered an accomplishment for an ROV to open and close valve handles on offshore platforms, new tool development now allows undersea vehicles to perform increasingly complex tasks in increasingly deeper water, including lubricating, pipe-cutting, making and breaking hydraulic connections, rigging support, and maintaining communications cable under water (Bannon, 1992; Gray et al., 1992).

Important tasks performed at mid-ocean depth, where capturing an object involves moving and controlling the vehicle and its manipulators as a single system, require more advanced capability. The new capability, "object-based task-level control," enables the human to command the task to be done; the control system then plans and executes the task, using the onboard vehicle manipulator control system. This capability will allow near real-time control of AUVs in midwater tasks (Wang et al., 1993).

Scientific applications also require manipulators and tools, especially for selecting and gathering samples. New and diverse tools being developed for these tasks will continue to increase the scope of scientific work that undersea vehicles can perform. Nonetheless, general-purpose manipulators will always be important in undersea vehicle science. Current low-cost ROVs usually come with some basic manipulation capabilities, and their dexterity will improve to approach that of the larger systems, although their load capability will likely remain limited (Schloerb, 1992; Sprunk et al., 1993). Vehicle manipulators can be improved and more sophisticated tasks commanded at the task level with the development of new underwater sensors for proximity, force, touch, and audio—to give the operator feedback on the performance of manipulators and other mechanical systems—which will most likely be based on devices created for terrestrial and space applications.

The payload-carrying capabilities of vehicles are important for selecting or designing tool packages. More capacity means heavier tools capable of performing greater work; and, especially in geology, carrying capacity affects the ability to transport sample rocks to the surface.

More research and development needs to be done before AUVs will be able to perform more than simple tasks. However, there are several applications that could use even primitive autonomous recognition and manipulation. This is particularly important when the surrounding geometry can be controlled and anticipated, such as maintaining structures whose physical characteristics are well known or structures specially built for robotic maintenance. Work in outer space has had to deal with similar time delays and task structure problems to those that are faced in some AUV designs. Therefore, it is expected that useful techniques for accomplishing tasks will derive from space-related work as well as from subsea contexts (JPL, 1995). Projects to develop robotics applicable to undersea vehicles have been completed or are under way at NASA's AMES facility, the Stanford University and MBARI joint program (Wang et al., 1993), and the Pennsylvania State University and Woods Hole Oceanographic Institution joint program.

Sensors

An important undersea vehicle mission is collecting data from various types of sensors. Thus, sensors tend to be a technology limiter or driver for vehicle applications. Sensors in the context of "payloads" refer to those sensors that are not directly involved in the functioning of the undersea vehicle but are used to collect data from various external sources, such as the environment. Such sensors can be carried by all classes of vehicles; typically, the sensors are matched to the type of host vehicle that is transporting and supporting them. For example, sensors that are applicable to large area searches would not normally be installed on DSVs, which may have poor range and endurance capabilities. Furthermore, the characteristics of specific vehicle types can have a significant influence on the design of sensors. A case in point is AUVs, where sensors are critical to overall capability, particularly because of limitations to direct interaction by humans in system control. From a handling and cost point of view, a common desire is to make AUVs as small as possible, thus imposing payload-carrying-capacity and resident-energy limitations. This, in turn, imposes similar restrictions on allowable sensors, which must be smaller and more energy efficient. If AUVs make possible longer missions than those of DSVs and ROVs, they will need sensors that are more resistant to fouling and with longer-lasting calibration characteristics. DSVs and ROVs will derive benefits from improvements in these same sensor characteristics.

Foreign work on sensors is exemplified by the European Community Cooperative Research Program, Marine, Science and Technology, which funds research and development in sampling and measuring instrumentation, including optical plankton analysis systems, electrochemical instrumentation for in situ determination of trace metals, in situ acoustic characterization of suspended sediment, and antifouling coatings for submarine sensors. Acoustic and optical sensors are the most broadly applicable and deserve the highest research and development priority.

Acoustic Sensors. Acoustic payload sensors include side-scan sonars and special scanning sonars, sub-bottom profilers, and imaging sonars. Side-scan sonars represent a well-developed engineering practice, with advances primarily in the areas of processing and display, not the onboard transceiver. However, in the case of AUVs, significant efforts have been directed toward digitizing and processing side-scan sonar signals on board the vehicle, either to conserve data storage space or to form the basis for autonomous action. Various types of scanning sonars can require significant onboard processing and may even contain their own processors. Synthetic aperture sonars, which promise to increase range and resolution by an order of magnitude, require very accurate navigation for short periods. If such sonars can be reduced in size and power consumption to be readily installed on AUVs, while at the same time meeting their navigation requirements, then all potential host vehicles will benefit. Acoustic imaging sensors have been under sporadic development for over 20 years. These sensors, which are useful even in murky water, have relatively high frequencies—on the order of 0.3–2.0 MHz—and can achieve image-level resolutions. Consequently, useful ranges are commonly limited to 100 meters or less. Scanning sensors have frequently been used on ROV systems for approximately 15 years. With a size reduction, these sensors can be installed and operated on some AUVs.

The trend toward processing acoustic data on board as a director for vehicle action, without human intervention, has several direct benefits. One is to reduce time required to implement such action; another may be to assist a human operator. Algorithms imbedded in resident software can assist with specific object recognition for tracking animals or locating objects. This can be done by programming the resident acoustic sensor processor with the unique characteristics of objects to be recognized. When these characteristics are matched with the incoming sensor data, a specific action, such as a vehicle maneuver or an alarm, results.

A variety of acoustic sensors are available to study the distribution and abundance of animals in the water column.

Currently, there are three principal architectures used to count targets, estimate target strength, and estimate volume backscattering. Multifrequency systems use inverse techniques to estimate biological properties; dual-beam and split-beam echo sounders measure the properties directly. Recent research has demonstrated wide variation in scattering properties of zooplankton and micronekton (Greene et al., 1989, 1991, 1994; Wiebe et al., 1990). Developments in these systems will be vital to making accurate estimates of biological properties.

Optical Sensors. Video cameras are the most commonly used optical sensors. A number of camera technologies, including charge-coupled devices (CCD) with one to three chips, silicon intensified target, low-light-level imaging tubes, stereo pair imaging, and CCD-based electronic still cameras, have been adapted from their parent commercial applications for use on undersea vehicles. Sources for these technologies include the television industry and military systems. Because of the severe limitations on available energy in AUVs, low-light imaging sensors (which do not require power-consuming lights) are an important optical payload component.

Video cameras and high-speed strobes can be configured to provide high-resolution images of plankton at frame rates of up to 60 per second. Individual targets as small as 60 microns can be identified. Automated image analysis techniques are being developed, but considerable work remains to be done in this area. Sample volumes at the highest resolution are extremely small, and much higher capacity CCDs are required. Increased image sizes, however, will require new data compression schemes that reduce the data loading and significantly enhance low-power data storage media.

As with acoustic perception capability, the real power of an artificial vision system is the perception capability that carries far beyond optical sensing per se. Providing a degree of scene analysis onboard the AUV makes it possible to send key information over the acoustic data link to the human task director in near real time, which in turn enables the human to direct tasks in near real time. For example, using newly developed bottom-mosaicing techniques, the AUV can accomplish local real-time vehicle guidance by scene matching. The scientist looking at the transmitted scene image can instruct the AUV to "hover over that starfish there," and the AUV will obey.

Other optical sensors that are emerging in development with direct applicability to undersea vehicles are the laser line scanner (LLS) images and range-gated laser imagers. LLS sensors work on the principle of a single laser, directed to scan a sector of the ocean bottom while the host vehicle moves forward at a uniform rate. The result is a good resolution waterfall (continuously scrolling) image display of much larger areas than conventional camera systems can cover in water of equivalent clarity. Recent enabling developments in LLS technology include the incorporation of solid-state lasers that require only 100 watts power and folded optics, that can reduce overall size by 50 percent. High sensor cost has restricted widespread use of the LLS to date.

Laser range-gating uses a traditional CCD image sensor, but "gates" the light reflected from a laser pulse to eliminate unwanted backscatter (Swartz, 1993). Unlike the LLS, which requires the vehicle to be moving to provide the second display axis, range-gated systems can be stationary, that permits precise positioning in the vicinity of an area of interest. Range-gated optical sensors have recently become available, and they have already undergone a tenfold reduction in required power and a significant size reduction; the cost remains high, however.

Chemical Sensors. Chemical sensors allow undersea vehicles to perform tasks that would otherwise require collecting water samples for laboratory analysis. It is necessary to distinguish between chemical sensors (devices in which passive diffusion transports the substance to be analyzed to the detector) and chemical analyzers (devices that actively transport the substance through an analytic process). Both types of devices can work on undersea vehicles.

Sensors are used to monitor dissolved chemicals. However, few chemical sensors with adequate sensitivity and selectivity are available today for the in situ determination of dissolved chemicals in seawater. Electrochemical sensors for oxygen and pH are the only devices that are in widespread use for in situ measurements. Changes in oxygen and pH reflect the rates of primary production (or respiration) through production of oxygen and consumption of carbon dioxide during photosynthesis. They do not provide direct information on the nutrient elements that limit these rates. Today, the only method for remotely monitoring dissolved nutrient or trace element (e.g., NO_3, NH_4, PO_4, SiO_2, Fe, Co, Mn, Zn) concentrations is to use fully automated chemical analyzers (devices in which mass transport moves the chemicals through the instrument) that are adapted to operate in situ. Work on such devices is in progress at several laboratories, and it is now possible to monitor concentrations of nutrient species for extended periods of time (up to several months) with sensors appropriate for mounting on a vehicle. Such analyzers are based on the principle of flow injection and are inherently more accurate and complex than sensor systems. Recent advances have made it possible to produce systems with only a few moving parts (Johnson et al., 1986b; Jannasch, 1992). Analyzers also have the advantage of providing a direct, chemical calibration while operating in situ. Chemical analyzers with sufficiently rapid response rates (< 30 seconds) for use on ROVs have been used in situ to map distributions of nitrate, silicate, sulfide, manganese, and iron (Johnson et al., 1986a; Johnson et al., 1990; Coale et al., 1991).

Currently, these analyzers have a lifetime of about one day when operated continuously, but their lifetime may be extended if operation is intermittent. They also are susceptible to clogging and to failures of valves, pumps, and other hardware. Commercial versions of these instruments are beginning to become available and should be accessible for deployment on operational vehicles in the near future. Expected development includes stop-flow systems, which can be activated at any time from a dormant state (Jannasch, 1992; NRC, 1993). Significant development also is likely to come from fiberoptic sensors ("optrodes"), light transmitted to a target through an optical fiber where it interacts with an indicator designed to respond to the presence, absence, or concentration of the substance to be analyzed. These sensors may be smaller, may require lower power, and may retain more stable calibrations than other types of sensors (Walt, 1992).

Conductivity, Temperature, Depth Sensors. Conductivity, temperature, and depth are among the most common sensors in oceanography and have undergone a great deal of technical development. The state of the art in CTDs has advanced to the point that commercial, off-the-shelf CTD sensors are acceptable for most situations encountered by undersea vehicles, including AUV requirements. Fouling of electrodes can be a problem for long-term deployments in shallow water (Bales and Levine, 1994), and new CTD developments focus on ultra-low-power specialized electronics to eliminate fouling and on dynamic calibration (Brown, 1991).

Fluorometers and Transmissometers. Fluorometers measure chlorophyll concentration in situ by strobing blue light (typically 480 nm) through a small volume of water (Bartz et al., 1988). Photons that enter the chloroplasts in phytoplankton cells cause the chlorophyll to emit light at a different wavelength (about 670 nm). This fluorescence is measured and, with proper calibration, can be related to the amount of chlorophyll in the volume. Typically, fluorometers consume 3 to 10 watts of power, weigh under 10 kg in air, and provide resolution of 0.01 mg/l for concentrations up to 100 mg/l.

Transmissometers measure particle density by projecting light of a particular wavelength (usually around 670 nautical miles) through a volume of water and measuring the amount of light received at the end of the path (Bartz et al., 1978). The decrease in light received relative to that transmitted is due to absorption of light by the water, dissolved organic matter and particulate matter, and scattering by particles in the water. With appropriate calibration, particle concentrations can be estimated (Bishop, 1986). Transmissometers tend to be lightweight and low power (0.1 to 0.3 w) but are rather large, because of the length of the light path (25 to 100 cm) through the sample volume. Replacing transmissometers with smaller backscatter detectors has been studied. This would provide a similar measurement, but in a size suitable for an AUV (Bales and Levine, 1994). The backscatter detector projects light into a sample volume using two modulated 880-nm light-emitting diodes, and the scattered light from particulate matter is detected by a "solar-blind" sensor.

The device weighs 0.26 kg in air and requires about 28 milliamperes at 9 to 28 volts. Calibration must be exact if more than a qualitative index of suspended particle concentration is needed. It is important however, that sensors on AUVs be equivalent to those currently used by scientists in other oceanographic applications.

Newer optical instrumentation has been developed that is designed to measure spectral absorption and attenuation simultaneously (Moore, 1994). The device requires a pumping system to continuously move water through two flow tubes: one to measure scattering and the other to measure both scattering and absorption. With data on up to nine wavelengths, it is possible to derive estimates of dissolved organic matter concentration, phytoplankton pigment type, and total chlorophyll content. The device weighs 7 kg in air, is about 70 cm long, and requires 8 w at 12 volts. Other upwelling and downwelling light sensors are also available, some of which would be useful in AUV applications.

Magnetic Field Sensors. Small flux-gate compasses are available at low cost for magnetic surveys. However, scientists also need total field measurements, which cannot be performed with these devices. Proton precession magnetometers are also available. These have been used since the mid-1960s for surface-based surveys (Larson and Spiess, 1969) and could be adapted to AUV applications (Tivey, 1992).

Gravity Sensors. The principal development work on compact gravimeters systems was done to support military objectives. Early models were used on DSVs in the 1960s. Smaller gravimeters are being developed that could be used on both ROVs and AUVs; however, accurate compact systems are still expensive and complex. Gravimeters are evolving rapidly.

Current Meters. Several sophisticated current meters are available, most of which could be adapted for undersea vehicle use. Vector current meters have been used in moorings to measure two or three components of current at a point as a function of time. Acoustic Doppler meters have also been used on ships and moorings to measure three components of current close to a platform using Doppler shift on backscattered sound pulses. With some modifications, these are being applied to vehicle navigation systems. Several new oceanographic current meters operate on principles that could be incorporated into miniaturized sensors suitable for installation on undersea vehicles.

Physical Samplers

In addition to data obtained by the various types of sensors and from onboard analytical devices, in situ physical samples need to be collected. As noted earlier, manipulators will remain the primary means for recovering samples from the seafloor. This includes direct manipulator intervention as well as the use of specialized tools, such as sediment corers, grab samplers, scoops, and sieving apparatuses. The ability to drill and recover rock cores from DSVs was demonstrated during *Alvin* dives in 1991 in the Juan de Fuca Ridge (Stakes et al., 1992). A core drilling system was added to the payload package of the ROV *Ventana* in 1992 (Stakes, 1996). Sampling devices used in the water column include pump-driven suction samplers, closing chambers, and "water catchers" (Youngbluth, 1984; Robison, 1993). For many biological, geological, and chemical samples, maintaining ambient temperature and pressure during transport to the surface is an important requirement, and several such systems are under development.

Efficient, large-capacity external stowage systems for physical samples are also an important part of the submersible vehicle's external payload. Generally these consist of bins, with or without lids, or hydraulically activated drawers that withdraw into the vehicle's framework. Elevator systems are being developed that can raise very large or heavy samples that exceed the vehicle's lifting payload to the surface independently.

Surface and Shoreline Support

A significant component of vehicle system technology and management is the surface support and the shore-side support, as shown in the schematic drawing in Figure 2-1. Surface support requirements vary widely among vehicle systems according to their size, characteristics, and the nature of their applications. Cranes and A-frames are the most common lifting elements, with the largest most complex vehicles requiring dedicated support vessels. Many modern systems are readily deployed from ships of opportunity. Both DSVs and AUVs require battery-charging and support facilities, and DSVs require recharging life support systems. ROVs need tether management and handling systems. In some ROV applications requiring precise placement, the surface support ships must have dynamic positioning capabilities.

Shore-side support is equally important to mission planning and development; storage, overhaul, and maintenance; and timely, accurate resupply. Training, while noted as a key element in shoreside support, is also at the heart of safe and successful vehicle system operations. Training is a key consideration when selecting the type of vehicle to perform a task.

Launch and Recovery

At-sea vehicle operations are generally limited by the sea state in which they can be launched and recovered. If the launch and recovery system can be designed to work safely in higher sea states, then the on-station time of the support vessel increases, and the vehicle is made more productive, that is, more diving days are available without shutting down operations because of weather and sea state.

Undersea vehicles are usually positioned navigationally and launched, tracked, and recovered by surface ships, semi-submersibles, or platforms. A-frames, elevators, and ramps are typically used to accomplish the recovery. Some vehicles use dedicated vessels and others can work from ships of opportunity. The size and the specialized handling equipment required for a particular vehicle, combined with the size of the crew required to support the operation, determine the size of the ship required for a particular mission and significantly contribute to overall mission cost.

Techniques for launching DSVs and ROVs are adequate for most vehicle applications; improvements in launching and retrieval are likely to have an incremental influence on overall system performance. An exception is research related to mitigating snap loads in long ROV umbilical lines when the vehicle is being deployed from a small vessel. Most large ROV launch and recovery systems have a motion-compensation and tether-compliance component to lessen the influence of sea surface motion on the vehicle. On the other hand, launch and recovery techniques for AUVs are still evolving. To take advantage of an AUV's lower cost and minimize surface support, new and unique launch and recovery techniques will be required.

SYSTEMS INTEGRATION

The systems engineering and integration process applies a formal, disciplined approach to focusing, with quantitative specifications (metrics), on the mission the system is to perform and on using the technologies, hardware, and software that will produce a harmonious union of subsystems to achieve the best combination of reliability, economy, and operational effectiveness. For the development of undersea vehicles this means the systematic integration of all the components of a vehicle system such that an optimum whole is achieved with respect to cost and effectiveness in performing task or mission objectives. As the engineering of submersible systems becomes more complex, the process necessarily becomes more formal and rigorous.

Systems engineering and integration (SE&I) is particularly relevant to undersea vehicles because of the diversity of the components and technologies involved. Some of the very elements that appear to be holding up wider usage of these systems are elements that lend themselves to SE&I. Examples are the integration of new sensors and other payloads with existing vehicles and the innovative merging of payload subsystems with new vehicles, both of which require careful attention to ensure reasonable cost, flexibility, and operational effectiveness.

In reviewing the preceding sections, it can be argued that a great deal of the current effort is ongoing with respect to components and operational techniques, but relatively little attention is being directed toward bringing all this technology together to form specific payloads and integrate them with new or existing vehicles in a cost-effective manner. In other words, new technology per se is only one enabling factor in achieving more widespread use of vehicles; the other is more effective system integration of the technologies that are already available. Implicit within this observation is the recognized value of using common interfaces, particularly mechanical, power, and signal interfaces, between payload components and host vehicles.

TECHNOLOGY TRANSFER FROM OTHER INDUSTRIES AND TECHNICAL FIELDS

Undersea vehicle development has drawn and will continue to draw on the technology bases from other industries and technical fields. Without this technology infusion, the cost of vehicle development would be prohibitively high, system support would be expensive, and rapid prototyping would be virtually impossible. Key examples of such technology transfer are summarized in Table 2-4. Note that for purposes of clarity, no distinction has been made between any of the undersea vehicle types (i.e., DSVs, ROVs, and AUVs). The table also includes the special requirements of undersea vehicle technology, in some cases calling for significant adaptations of components transferred from other fields.

Since so much of the technology for undersea vehicles comes from other fields, the question then becomes how best to take advantage of that transfer. Rarely is the technology automatically transferable and usable without cost. Rather, technology transfer occurs through focused efforts, usually as communications among technical experts. In addition, the movement of skilled technical leaders from one industrial or technical sector to another provides an efficient mechanism for technology transfer. Such efforts tend to be inexpensive and can have a great impact if they improve the total level of transfer by even a small percentage.

Still, the amount of effort required for technology transfer is proportional to the advantages it yields and the level at which it occurs. High-level ideas from other disciplines can be the most effective contributors to technology transfer, but they also require the most work to adapt and implement. It is wrong to assume that because technologies are under development in other technical disciplines they require no development within the field of undersea vehicles. The process of adaptation alone imposes significant effort and cost. For a technology assessment, it is worth asking whether one can anticipate developments in undersea vehicles by looking at the future (or even current) directions of these other disciplines. It is probable, for example, that significant progress will be made in energy systems that can be applied to undersea vehicles (particularly AUVs) because of the development effort currently focused on power systems for electric cars. Similarly, as defense-dependent industries, such as aerospace, feel the effects of the current reduction in defense budgets, they will seek new markets for the technologies developed for military applications. The undersea vehicle

TABLE 2-4 Technology Transfer

Vehicle Subsystem	Other Industries and Disciplines	Unique Requirements and Adaptations for Undersea Vehicles
Energy	Auto industry/electric cars, computers, and communications	Air independence, shipboard handling
Propulsion	Hydraulics, pumps, motors, valves, filters, plumbing, brushless dc motors, propellers	Hydrodynamics, pressure tolerance, ability to work in oil
Materials and Structures	Aerospace, boat building, aluminum composites, 316 SS, acrylics, graphite reinforced plastics	Pressure tolerance, corrosion resistance
Navigation and Positioning	Aerospace/compass and gyros, video cameras, lighting, global positioning system/inertial navigation system (GPS/INS)	Need to operate in acoustic rather than radio regime; GPS available only occasionally
Guidance and Mission Control	PC industry, automatic control	Unique hydrodynamics, long-term reliability
Data Processing	PC industry, object-oriented programming, computer-aided software engineering, computer science	Packaging for pressure housings, uniqueness of acoustic signal processing, pressure-tolerant electronics
Communications	Fiber optics, signal processing, electronics	Electromagnetic spectrum not available, acoustic medium only; packaging space restrictions
Task-Performance Systems and Tools	Construction, robotics, and automation	Moving platform and manipulator system, acoustic bandwidth, denser medium, high pressure
Sensors	Other ocean sciences, instrumentation, micromachinery, medical sensors	Seawater medium, long-term stability, biological fouling, corrosion
Launch and Recovery	Other marine applications/boat handling	Ability to work in multiple sea states, tether handling

engineering community can take advantage of these trends (e.g., laser gyroscopes and inertial navigation systems) by remaining vigilant about how its own needs overlap with those of other industries.

FINDINGS

Finding. Research and development programs outside the field of undersea vehicles have had major impacts on undersea vehicle technology (Table 2-4). The automobile industry has led the development of batteries; the computer industry, the development of onboard processing; and the communications industry, the development of fiber-optic cable and signal processing techniques (although applying those developments in the ocean environment has required considerable effort). However, the few important subsystems that are unique to undersea vehicles, such as launch and recovery, small acoustic sensors, and undersea vehicle and manipulator control systems that can be commanded acoustically at the task level, do not receive attention from other industries and will progress only from well-supported and sustained efforts of the undersea community itself.

Finding. The subsystem technologies that are essential to highly effective undersea vehicle systems are in various states of evolution. These technologies include the following:

- **Energy.** A great deal of development effort in energy sources has taken place outside the undersea vehicle field. ARPA has supported work fuel cells, specifically for undersea vehicles, and progress is being made. However, there is a need for high-energy-density, low-cost-energy sources that can be commonly used on vehicle systems and that are not under development elsewhere. These energy sources are essential for operation of untethered AUVs, as well as AUVs used in the hybrid mode with telemetry tethers, to accommodate operations requiring long periods of time and long-distance traverses, such as large-area surveys.
- **Propulsion.** Some interesting developments have taken place, but current technology is generally satisfactory for most foreseeable requirements.
- **Navigation and Positioning.** Several acoustic systems are available for relative navigation over small to medium distances (less than 3.5 km, or 2 nmi), using local networks of transmitters with accuracies in the range of centimeters. Dead reckoning capabilities have evolved, fusing information from multiple sensors for increased accuracy. Positioning over longer times and distances, however, will continue to require the use of geodetic coordinates. Locating surface ships and buoys with very high precision using differential Global Positioning System measurements is commonly accomplished at low cost.
- **Guidance and Mission Control.** Guidance and control of ROVs have been improved through automation. Even greater potential returns are available from automation of AUVs, but unique requirements of undersea vehicles will necessitate major advances in control logic to permit widespread practical use of AUVs beyond the simple tasks that are currently automated. Moreover, new advances in task-level control architecture and acoustic bandwidth will permit human operators to direct AUVs, observe events, and issue high-level commands in real time—a major advance in the range of tasks that AUVs can perform.
- **Materials and Structures.** The current state of practice is generally satisfactory for most applications of undersea vehicles. Major benefit will come from the ceramics work initiated by the Navy for lightweight, deep-ocean, pressure-resistant housings and buoyancy structures. Pressure-tolerant electronics may allow significant weight reductions with the elimination of some pressure housings, but the cost of providing these changes may offset the limited weight reductions achievable.
- **Communication.** Significant advances in wideband acoustic communications are being made, with parallel advances in computing power. This, and the development of task-level controls, are the pair of technologies that will enable the powerful human near real-time command of AUVs. Work now being supported by ARPA and the Navy is beginning to provide good short-range communications for AUVs. Using optic fibers in ROV tethers has increased the bandwidth and enhanced the supervisory control capability of those systems.
- **Data Processing.** Computational power is linked directly to developments in the computer industry. Issues for undersea vehicles are strongly oriented to AUV onboard data logging and manipulation, in situ analyses, and systems management. These will involve low power consumption, high-volume data storage, and innovative manipulation techniques. As with guidance and control, the state of practice is embryonic, but significant advances in adaptations of data processing to undersea vehicles will have a major impact on the practicality of AUV operations.
- **Launch and Recovery.** There appear to be no serious technology issues here. This area is constantly evolving, with most emphasis on reducing size so that smaller, lower-cost surface support vessels can be used.
- **Task-Performance Control Systems.** General-purpose manipulators and tools are reasonably well developed for most anticipated tasks. Improved dexterity can be achieved for special purposes by adapting industrial robot manipulators, which have sensors for proximity, force, and touch. New real-time task planning and task-management architecture with sophisticated interfaces will enable humans to direct operations from the task

level for AUVs equipped with the expanded bandwidth now being developed.

- **Sensors.** Sensor capabilities impose limits on undersea vehicle missions. A wide variety of ocean-specific sensors is needed—acoustic, optical, chemical, magnetic, and others—to improve the abilities of undersea vehicles to search, identify, and measure pollutants and other vital substances; to help make undersea work more dexterous; and to present human operators with visual and other information on undersea activities. Advances in optical and acoustic sensor systems have been made, but much remains to be done, especially in the area of chemical sensors. The integration or fusion of data from different sensors offers significant potential for advancing this field.

Finding. Foreign developments in undersea vehicle technologies can contribute to U.S. needs. By being well aware of technical developments in foreign countries and participating in selected projects, researchers in the United States can incorporate that knowledge into their own work, thus avoiding duplication of research. The United States participates in cooperative development or bilateral exchange programs with a number of other countries, including China, France, and Japan.

REFERENCES

Apel, J. 1993. Norwegian Defense Research Establishment. Mg-Seawater-Battery-Powered Autonomous Underwater Vehicle, January–September 1992. Circulated memo. Laurel, Maryland: The Johns Hopkins Applied Physics Laboratory.

Asakawa, K., J. Kojima, Y. Ito, Y. Shirasaki, and N. Kato. 1993. Development of autonomous underwater vehicle for inspection of underwater cables. Pp. 208–216 in Proceedings, Underwater Intervention '93 held January 18–21, 1993 in New Orleans, Louisiana. The Marine Technology Society and the Association of Diving Contractors. Washington, D.C.: Marine Technology Society.

Ashley, S. 1993. Voyage to the bottom of the sea. Mechanical Engineering 115(12):50–58.

Bales, J.W., and E.R. Levine. 1994. Sensors for oceanographic applications of autonomous underwater vehicles. Pp. 434–446 in Proceedings Manual, AUVS '94 Technical Symposium held May 23–25, 1994 in Detroit, Michigan. MIT Sea Grant Report 94-26J. Arlington, Virginia: Association of Unmanned Vehicle Systems International.

Bannon, R.T. 1992. Deep recovery and inspection—Advanced SCARAB systems. Pp. 221–227 in Proceedings of the 10th Annual Conference, Intervention/ROV '92 held June 10–12, 1992 in San Diego, California. Washington, D.C.: Marine Technology Society.

Bartz, R., J.R.V. Zaneveld, and H. Pak. 1978. A transmissometer for profiling and moored observations in water. Pp. 102–108 in Proceedings of the Society of Photo-Optical Instrumentation Engineers, Ocean Optics (V) held August 30–31, 1978 in San Diego, California. M.B. White and R.E. Stevenson, eds. Bellingham, Washington: Society of Photo-Optical Instrumentation Engineers.

Bartz, R., R.W. Spinrad, and J.C. Glizard. 1988. A low power, high resolution, in situ fluorometer for profiling and moored observations in water. Pp. 157–170 in Proceedings of the Society of Photo-Optical Instrumentation Engineers, Ocean Optics (IX) held April 4–6, 1988 in Orlando, Florida. Bellingham, Washington: Society of Photo-Optical Instrumentation Engineers.

Bellingham, J.G. 1995. Personal communication to Donald W. Perkins, July 14, 1995. National Research Council. Washington, D.C.: Marine Board.

Bellingham, J.G., C.A. Goudey, T.R. Consi, and C. Chryssostomidis. 1992. A small, long range autonomous vehicle for deep ocean exploration. Pp. 461–467 in Proceedings of the 2nd International Offshore and Polar Engineering Conference held June 14–19, 1992 in San Francisco, California. MIT Sea Grant Report 93-18J. Cambridge, Massachusetts: MIT.

Bellingham, J.G., C.A. Goudey, T.R. Consi, J.W. Bales, D.K. Atwood, J.J. Leonard, and C. Chryssostomidis. 1994. A second generation survey AUV. Pp. 148–155 in Proceedings of IEEE AUV '94 held July 19–20, 1994 in Cambridge, Massachusetts. MIT Sea Grant Report 94-25J. Cambridge, Massachusetts: MIT.

Bellingham, J.G., M. Deffenbaugh, J.J. Leonard, J. Catipovic, and H. Schmidt. 1993. Arctic under-ice survey operations. Pp. 50–59 in Proceedings of the 8th International Symposium on Unmanned Untethered Submersible Technology held September 27–29, 1993 at the University of New Hampshire, Durham. Document Number 93-9-01. Lee, New Hampshire: Autonomous Undersea Systems Institute.

Bellingham, J.G., and J.J. Leonard. 1994. Task configuration with layered control. Pp. 193–202 in Proceedings of Mobile Robots for SubSea Environment, International Advanced Robotics Programme (IARP) held June 3–6, 1994 in Monterey, California. MIT Sea Grant Report 94-24J. Pacific Grove, California: Monterey Aquarium Research Institute.

Bishop, J.K.B. 1986. The correction and suspended particulate matter calibration of Sea Tech transmissometer data. Deep-Sea Research 33: 121–134.

Blase, E.F., and R.F. Bis. 1990. Power source selection for operation at deepest ocean depths. Marine Technology Society Journal 24(2):63–66.

Bowen, A.D., and B.B. Walden. 1993. Manned versus unmanned: A complementary approach. Marine Technology Society Journal(Winter): 92–93.

Bradley, A., and D.R. Yoerger. 1993. Design and testing of the Autonomous Benthic Explorer. Pp. 1044–1055 in Proceedings of the 20th Annual Symposium of the Association of Unmanned Vehicle Systems held June 28–30, 1993 in Washington, D.C. Arlington, Virginia: Association of Unmanned Vehicle Systems International.

Brininstool, M.R., and J.H. Dombrowski. 1992. NRAD undersea fiber-optic development and technology transfer. Pp. 473–478 in Proceedings of the 10th Annual Conference, Intervention/ROV '92 held June 10–12, 1992 in San Diego, California. Washington, D.C.: Marine Technology Society.

Broad, W.J. 1993. Racing to the bottom of the sea. New York Times, August 3:C1.

Brown, N. 1991. The history of salinometers and CTD sensor systems. Oceanus 34(1):61–66.

Catipovic, J. 1990. Performance limitations in acoustic telemetry. IEEE Journal of Ocean Engineering 15:205–216.

Catipovic, J. 1995. Personal communication to Donald W. Perkins, November 6, 1995.

Catipovic, J. 1996. Personal communication to Donald W. Perkins, August 6, 1996.

Coale, K.H., C.S. Chin, G.J. Massoth, K.S. Johnson, and E.T. Baker. 1991. In situ chemical mapping of dissolved iron and manganese in hydrothermal plumes. Nature 352:325–328.

Collins, K. 1993. Cost-effective AUVs for today's offshore industry. Pp. 199–207 in Proceedings of the 11th Annual Conference, Underwater Intervention held January 18–21, 1993 in New Orleans, Louisiana. Washington, D.C.: Marine Technology Society.

Collins, K., J. Stannard, R. Dubois, and G. Scamans. 1993. An aluminum-oxygen fuel cell power system (fcps) for underwater vehicles. Pp. 199–207 in Proceedings of the 11th Annual Conference, Underwater Intervention '93 held January 18–21, 1993 in New Orleans, Louisiana. Washington, D.C.: Marine Technology Society.

Curtin, T.B., J.G. Bellingham, J. Catipovic, and D. Webb. 1993. Autonomous oceanographic sampling networks. Oceanography 6(3):86–94.

DeRoos, B.G., C.R. Miele, K.B. Scott, and J.P. Downing. 1993. Deep ocean alumina ceramic pressure housing design and testing. Pp. 225–232 in Proceedings of the 11th Annual Conference, Underwater Intervention '93 held January 18–21, 1993 in New Orleans, Louisiana. Washington, D.C.: Marine Technology Society.

Ezekiel, T. 1991. Recent developments in optical gyros for inertial navigation. P. 7 in Sensor and Navigation Issues for Unmanned Underwater Vehicles, Moore, J., Jr., ed. MIT Marine Industry Collegium. MIT Sea Grant Report 90–26. Cambridge, Massachusetts: MIT Marine Industry Collegium.

Fossen, T.I. 1994. Guidance and Control of Ocean Vehicles. New York, New York: John Wiley & Sons.

Fricke, J.R. 1992. Applications of underwater vehicles to the offshore oil and gas industry. MIT Sea Grant College Program's Marine Industry Collegium and C.S. Draper Laboratories. Pp. 45–47 in Proceedings of the Workshop on Scientific and Environmental Data Collection with Autonomous Underwater Vehicles held March 3–4, 1992 in Cambridge, Massachusetts. MIT Sea Grant Report 92–2. Cambridge, Massachusetts: MIT Sea Grant Program.

Gangadharan, S., and H. Krein. 1989. Jet-propelled remote-operated underwater vehicles guided by tilting nozzles. Marine Technology 26:131–144.

Gentry, L.L. 1995. Personal communication to Committee on Undersea Vehicles and National Needs, August 18, 1995.

Gibbons, D.W., M. Niksa, and S. Sinsabaugh. 1991. Aluminum/oxygen fuel cell with continuous electrolyte management. Pp. 487–502 in Proceedings 18th Annual AUVs Technological Symposium held August 13–15, 1991 in Cambridge, Massachusetts. Washington, D.C.: Association for Unmanned and Exhibit Vehicle Systems.

Gray, W.E., D.A. Gray, and W.M. McDonald. 1992. Wet tests, diverless deepwater tie-in system. Pp. 13–29 in Proceedings of the 10th Annual Conference, Intervention/ROV '92 held June 10–12, 1992 in San Diego, California. Washington, D.C.: Marine Technology Society.

Greene, C.H., P.H. Wiebe, and J.E. Zamon. 1989. Acoustic visualization of path dynamics ocean ecosystems. Oceanography 7:1–9.

Greene, C.H., T.K. Stanton, P.H. Wiebe, and S. McClatchie. 1991. Acoustic estimates of Antarctic drill. Nature 349:110.

Greene, C.H., P.H. Wiebe, and J. Burczynski. 1994. Analyzing zooplankton size distributions using high-frequency sound. Limnology and Oceanography 34:129–139.

Greig, A.R., Q. Wang, and D.R. Broome. 1992. Weld tracking with a robotic manipulator fitted with a complaint wrist unit. Pp. 310–324 in Proceedings of the 10th Annual Conference, Intervention/ROV '92 held June 10–12, 1992 in San Diego, California. Washington, D.C.: Marine Technology Society.

Grey, A.C. 1992. Dispensed fiber-optic capabilities for ROV control and data transmission. Pp. 461–472 in Proceedings of the 10th Annual Conference, Intervention/ROV '92 held June 10–12, 1992 in San Diego, California. Washington, D.C.: Marine Technology Society.

Gritton, B.R., and C.H. Baxter. 1993. Video database systems in the marine sciences. Marine Technology Society Journal 26(4):59–72.

Grose, B.L. 1991. The correlation sonar, an absolute velocity sensor for autonomous underwater vehicle navigation. MIT Sea Grant College Program, Industry Collegium held January 15–16, 1991 in Cambridge, Massachusetts. EDO Report No. 11189. Cambridge, Massachusetts: MIT.

Gwynne, O., C. Stoker, D. Barch, L. Richardson, and P. Ballou. 1992. Telepresence control of ROVs: Application to undersea and future space exploration. Pp. 102–107 in Proceedings of the 10th Annual Conference, Intervention/ROV '92 held June 10–12, 1992 in San Diego, California. Washington, D.C.: Marine Technology Society.

Harma, D.A. 1988. Suitability of silver-zinc and silver-cadmium electrochemistries to provide electrical power for undersea systems. Pp. 1–20 in the Proceedings of the AUV 15th Annual Symposium and Exhibition held June 6–8, 1988 in San Diego, California. Arlington, Virginia: Association of Unmanned Vehicle Systems International.

Hawkes, G.S., and P.J. Ballou. 1990. Ocean everest concept: A versatile manned submersible for full ocean depth. Marine Technology Society Journal 24(3):79–86.

Healy, A.J., and D. Leonard. 1993. Multivariable sliding mode control for autonomous diving and steering of unmanned underwater vehicles. IEEE Journal of Ocean Engineering 18(3):327–339.

Howland, J.C., M. Marra, D.F. Potter, and W.K. Stewart. 1993. Near-real-time GIS in deep-ocean exploration. Report No. WHOI-CONTRIB-8046. Springfield, Virginia: National Technical Information Service.

Hughes, T.G. 1995. Advanced Research Laboratories/Pennsylvania State University. Presentation to American Defense Preparedness Association, Undersea Warfare Systems Division Symposium held October 18, 1995 in Washington, D.C. (Presentation not published.)

Hutchison, B., and B. Skov. 1990. A system approach to navigating and piloting small unmanned underwater vehicles. Pp. 129–136 in Proceedings of Symposium on Autonomous Underwater Vehicle Technology held June 5–6, 1990 in Washington, D.C. Piscataway, New Jersey: IEEE.

Hutchison, B. 1991. Velocity-Aided Inertial Navigation Systems. Pp. 8-9 in Sensor and Navigation Issues for Unmanned Underwater Vehicles. J. Moore, Jr., ed. MIT Sea Grant Report 90–26. Cambridge, Massachusetts: MIT Marine Industry Collegium.

Jannasch, H.W. 1992. In situ Chemical Detectors for Potential use on Autonomous Underwater Vehicles. Pp. 41–44 in Scientific and Environmental Data Collection with Autonomous Underwater Vehicles. J. Moore, ed. MIT Sea Grant Report 92-2. Cambridge, Massachusetts: MIT Sea Grant Program.

JPL (Jet Propulsion Laboratory). 1995. Space technology underwater: Undersea technology at the jet propulsion laboratory. Pp. 1505–1510 in Volume 3, Proceedings of Oceans '95 held October 9–12, 1995 in San Diego, California. Washington, D.C.: Marine Technology Society, Washington, D.C.

Johnson, K.S., C.L. Beehler, and C.M. Sakamoto-Arnold. 1986a. A submersible flow analysis system. Analytica Chimica Acta 79:245–257.

Johnson, K.S., C.L. Beehler, C.M. Sakamoto-Arnold, and J. J. Childress. 1986b. In situ measurements of chemical distributions in a deep-sea hydrothermal vent field. Science 231:1139–1141.

Johnson, K.S., C.M. Sakamoto-Arnold, and C.L. Beehler. 1990. Continuous determination of nitrate concentrations in situ. Deep-Sea Research 36:1407–1413.

Kunzig, R. 1996. A thousand diving robots. Discover 17(4):60–71.

Kurkchubasche, R. 1992. Elastic stability considerations for deep submergence ceramic pressure housings. Pp. 143–150 in Proceedings of the 10th Annual Conference, Intervention/ROV '92 held June 10–12, 1992 in San Diego, California. Washington, D.C.: Marine Technology Society.

Langrock, D.G., P. Richards, and J.M. Howard. 1992. ROV intervention system for installation and maintenance of subsea oil wells. Pp. 1–12 in Proceedings of the 10th Annual Conference, Intervention/ROV '92 held June 10–12, 1992 in San Diego, California. Washington, D.C.: Marine Technology Society.

Larson, R.L., and F.N. Spiess. 1969. East Pacific Rise Crest: A near-bottom geophysical profile. Science 163:68–71.

Mackelburg, G.R. 1991. Acoustic data links for UUVs. Pp. 1400–1406 in Proceedings of the IEEE Oceans '91 Conference held October 1–3, 1991 in Honolulu, Hawaii. Piscataway, New Jersey: IEEE Service Center.

Marks, R.L., M.J. Lee, and S.M. Rock, 1994a. Visual sensing for control of an underwater robotic vehicle. Pp. 213–225 in Proceedings of the IARP 2nd Workshop on Mobile Robots for Subsea Environments held May 3–6, 1994 in Monterey, California. Pacific Grove, California: Monterey Bay Aquarium Research Institute.

Marks, R.L., H.H. Wang, M.J. Lee, and S.M. Rock. 1994b. Automatic visual station keeping of an underwater robot. Pp. 137–142 in Volume 2, Proceedings of IEEE Oceans '94 held September 13–16, 1994 in Brest, France. New York: IEEE.

McFarlane, J.R. 1987. The genesis and metamorphosis of underwater work vehicles. Pp. 115–127 in Undersea Teleoperators and Intelligent

Autonomous Vehicles, N. Doelling, and E. Harding, eds. MIT Sea Grant Report 87-1. Cambridge, Massachusetts: MIT Sea Grant College Program.

Meyer, A.P. 1993. Development of proton exchange membrane fuel cells for underwater applications. Pp. 146–151 in Proceedings of Oceans '93 held October 18–21, 1993 in Victoria, British Columbia, Canada. New York: IEEE.

Michel, J.L, T. Conway, and H. Le Roux. 1987. Epaulard: Operational developments. Pp. 14–17 in Proceedings of the 5th Annual International Symposium on Unmanned Untethered Submersibles Technology held June 22–24, 1987 at the University of New Hampshire, Durham. Durham, New Hampshire: University of New Hampshire.

Michel, J.L., and H. Le Roux. 1981. Epaulard: Deep bottom surveys now with acoustic remote controlled vehicle, first operational experience. Pp. 99–103 in the Proceedings of Oceans '81 held September 16–18, 1981 in Boston, Massachusetts. Washington, D.C.: Marine Technology Society.

Mooney, J.B., H. Ali, R. Blidberg, M.J. DeHaemer, L.L. Gentry, J. Moniz, and D. Walsh. 1996. World Technology Evaluation Center Program (WTEC). World Technology Evaluation Center Panel Report on Submersibles and Marine Technologies in Russia's Far East and Siberia. International Technology Research Institute, in press. Baltimore, Maryland: Loyola College of Maryland.

Moore, C. 1994. In-situ, biochemical, oceanic, optical meters. Sea Technology 35(2):10–16.

Moore, J., Jr. (ed.). 1988. Power Systems for Small Underwater Vehicles. MIT/Marine Industry Collegium Opportunity Brief. MIT Sea Grant Report 88-11. Cambridge, Massachusetts: MIT Sea Grant Program.

Moore, J., Jr. (ed.). 1991. Sensor and Navigation Issues for Unmanned Underwater Vehicles. MIT Sea Grant Collegium Opportunity Brief. MIT Sea Grant Report 90-26. Cambridge, Massachusetts: MIT Sea Grant Program.

Negahdaripour, S. 1993. Optical sensing for autonomous subsea vehicles. Pp. 10–12 in Perception, Scene Reconstruction, and World Modeling for Unmanned Underwater Vehicles. MIT Sea Grant Collegium Opportunity Brief. MIT Sea Grant Report 92-24. Cambridge, Massachusetts: MIT Sea Grant Program.

Newman, J.B., and B.H. Robison. 1993. Development of a dedicated ROV for ocean science. Marine Technology Society Journal 26(4):46–53.

NRC (National Research Council). 1993. Applications of Analytical Chemistry to Oceanic Carbon Cycle Studies. Committee on Oceanic Carbon, Ocean Studies Board, NRC. Washington, D.C.: National Academy Press.

Pappas, G. 1995. Personal communication to Donald W. Perkins, November 3, 1995.

Pappas, G., R. Rosenfeld, and A. Beam. 1993. The ARPA/Navy unmanned undersea vehicle program. Unmanned Systems 11(2).

Perrier, M., and J.G. Bellingham. 1992. Control Software for an Autonomous Survey Vehicle. MIT SEA Grant Report 93–20J. Cambridge, Massachusetts: MIT SEA Grant Program.

Ricks, D.C. 1989. A project to develop and test layered control systems for underwater vehicles. Pp. 123–127 in Proceedings of the 8th International Offshore Mechanics and Arctic Engineering Conference held March 13–23, 1989 in The Hague, Netherlands. New York: ASME.

Robison, B.H., K.R. Reisenbichler, and S.A. Etchemendy. 1992. A scientific perspective on the relative merits of manned and unmanned vehicles. Pp. 485–489 in Proceedings of the 10th Annual Conference, Intervention/ROV '92 held June 10–12, 1992 in San Diego, California. Washington, D.C.: Marine Technology Society.

Robison, B.H. 1993, Midwater research methods with MBARI's ROV. Marine Technology Society Journal 26(4):32-39.

Robison, B.H. 1994, New technologies for sanctuary research. Oceanus 36:75–80.

Rosenblum, L.J., W.K. Stewart, and B. Kamgar-Parti. 1993. Undersea visualization: A tool for scientific and engineering progress. Pp. 205–223 in Animation and Scientific Visualization Tools and Applications. Earnshaw, and Watson, eds. London: Academic Press.

Scherbatyuk, A. 1993. A side-scan sonar image processing system for the survey of pipeline. Pp. 68–75 in Proceedings of the 10th Annual Conference, Intervention/ROV '92 held June 10–12, 1992 in San Diego, California. Washington, D.C.: Marine Technology Society.

Schloerb, D.W. 1992. Development of a four-function mini-ROV manipulator for marine scientists. Pp. 207–220 in Proceedings of the 10th Annual Conference, Intervention/ROV '92 held June 10–12, 1992 in San Diego, California. MIT Sea Grant Report 93-2J. Washington, D.C.: Marine Technology Society.

Seymour, R.J., D.R. Blidberg, C.P. Brancart, L.L. Gentry, A.N. Kalvaitis, M.L. Lee, J.B. Mooney, and D. Walsh. 1994. World Technology Evaluation Center Program. Pp. 150–262 in World Technology Evaluation Center Panel Report on Research Submersibles and Undersea Technologies. NTIS Report No. PB94-184843. Baltimore, Maryland: Loyola College of Maryland.

Sloan, F., and H. Nguyen. 1992. Use of extended-chain polyethylene (ecpe) fibers in marine composite applications. Pp. 173–182 in Proceedings of the 10th Annual Conference, Intervention/ROV '92 held June 10–12, 1992 in San Diego, California. Washington, D.C.: Marine Technology Society.

Somers, T. and F. Geisel. 1992. Subsea dynamic positioning of ROVs. Pp. 369–373 in Proceedings of the 10th Annual Conference, Intervention/ROV '92 held June 10–12, 1992 in San Diego, California. Washington, D.C.: Marine Technology Society.

Sprunk, H.J, P.J. Auster, L.L. Stewart, D.A. Lovalvo, and D.H. Good. 1993. Modifications to low-cost remotely operated vehicles for scientific sampling. Marine Technology Society Journal 26(4):54-58.

Stachiw, J.D., and B. Frame. 1988. Graphite-Fiber-Reinforced Plastic Pressure Hull Mod 2 for the Advanced Unmanned Search System Vehicle. Technical Report NOSC No. 1245. San Diego, California: Naval Ocean Systems Center (now the Naval Command, Control, and Ocean Surveillance Center).

Stachiw, J.D. 1992. Engineering Criteria Used in the Selection of Ceramic Composition for External Pressure Housings. San Diego, California: Naval Command, Control, and Ocean Surveillance Center.

Stachiw, J.D. 1993. Quoted in Voyage to the Bottom of the Sea. Mechanical Engineering 115(12):56.

Stakes, D., W.S. Moore, T. Tengdin, H. Holloway, M. Tivey, M. Hannington, J. Edmond, and J.F. Todd. 1992. Core drilled into active smokers on Juan de Fuca Ridge. Transactions in the American Geophysical Union (EOS) 73(26):273, 278–279, 283.

Stakes, D. 1996. Personal communication to Donald W. Perkins, August 16, 1996.

Stannard, J.H., G.D. Deuchars, J.R. Hill, and D. Stockburger. 1995. Sea trials of an aluminum/hydrogen peroxide unmanned underwater vehicle propulsion system. Pp. 181–191 in Proceedings Manual, Technical Papers, AUVS '95 Conference held July 10–12, 1995 in Washington, D.C. Arlington, Virginia: Association of Unmanned Vehicle Systems International.

Stojanovic, M., J. Catipovic, and J. Proakis. 1993. Adaptive multichannel combining and equalization for underwater acoustic communication. Journal of Acoustical Society of America 94(3)(Part 1):1621–1631.

Stojanovic, M., J. Catipovic, and J. Proakis. 1995. Reduced complexity spatial and temporal processing of underwater acoustical command signals. Journal of Acoustical Society of America 98(2)(Part 1):961–972.

Stoker, C.R. 1994. From Antarctic to space: Use of telepresence and virtual reality in control of a remote underwater vehicle. MOBILE ROBOTS IX. Pp. 288-299 in Proceedings of SPIE held November 2–4, 1994 in Boston, Massachusetts. Bellingham, Washington: Society of Photo-Optical Instrumentation Engineers.

Stommel, H. 1989. The Slocum Mission. Oceanus 32(Winter 89/90): 93–96.

Sucato, P.J. 1993. Direct flowline pull-in and connection operations by ROV. Pp. 133–139 in Proceedings of the 11th Annual Conference,

Underwater Intervention '93 held January 18–21, 1993 in New Orleans, Louisiana. Washington, D.C.: Marine Technology Society.

Swartz, B.A. 1993. Diver and ROV deployable laser range gate underwater imaging systems. Pp. 193–198 in Proceedings of the 11th Annual Conference, Underwater Intervention '93 held January 18–21, 1993 in New Orleans, Louisiana. Washington, D.C.: Marine Technology Society.

Tivey, M.A. 1992. Micro-magnetic field measurements near the ocean floor. Pp. 49–52 in Scientific and Environmental Data Collection with Autonomous Underwater Vehicles. J. Moore, Jr., ed. MIT Sea Grant Report 92-2. Cambridge, Massachusetts: MIT Sea Grant Program.

Triantafyllou, M. 1992. Large-scale circulation studies with multiple underwater vehicles. Pp. 25–28 MIT Sea Grant College Program's Marine Industry Collegium and C.S. Draper Laboratories Workshop on Scientific and Environmental Data Collection with Autonomous Underwater Vehicles held March 3–4, 1992 in Cambridge, Massachusetts. MIT Sea Grant Report 92-2. Cambridge, Massachusetts: MIT Sea Grant Program.

Triantafyllou, M.S., G.S. Triantafyllou, and R. Gopalkrishnan. 1992. Wake mechanics for thrust generation in oscillating foils. Physics of Fluids 3(12):2835–2837.

Tusting, R.F., and D.L. Davis. 1993. Laser systems and structured illumination for quantitative undersea imaging. Marine Technology Society Journal 26(4):5–12.

Walsh, D. 1994. Undersea satellites: The commercialization of AUVs. Marine Technology Society Journal 27(4):54–63.

Walt, D.R. 1992. Recent developments and trends in fiber optic chemical sensors. Pp. 37-41 in Scientific and Environmental Data Collection with Autonomous Underwater Vehicles. J. Moore, ed. MIT Sea Grant Report 92-2. Cambridge, Massachusetts: MIT Sea Grant Program.

Walton, J.M. 1991. Advanced unmanned search system. Pp. 1392–1399 in Proceedings of the IEEE Oceans '91 Conference held October 1–3, 1991 in Honolulu, Hawaii. Piscataway, New Jersey: IEEE Service Center.

Walton, J., M. Cooke, and R. Uhrich. 1993. Pp. 243–249 in Proceedings of the 11th Annual Conference, Underwater Intervention '93 held January 18–21, 1993 in New Orleans, Louisiana. Washington, D.C.: Marine Technology Society.

Wang, H.H., R.L. Marks, S.M. Rock, M.J. Lee, and R.C. Burton, 1992. Combined camera and vehicle tracking of underwater objects. Pp. 325–332 in Proceedings of the 10th Annual Conference, Intervention/ROV '92 held June 10–12, 1992 in San Diego, California. Washington, D.C.: Marine Technology Society.

Wang, H.H., R.L. Marks, S.M. Rock, and M.J. Lee. 1993. Task-based control architecture for an untethered, unmanned, submersible. Pp. 131–147 in Proceedings of the 8th International Symposium on Unmanned Untethered Submersible Technology held September 27–29, 1993 at the University of New Hampshire, Durham. Document Number 93-9-01. Lee, New Hampshire: Autonomous Undersea Systems Institute.

Wang, H.H., R.L. Marks, T.W. McLean, S.D. Fleischer, D.W. Miles, G.A. Sapilewski, S.M. Rock, M.J. Lee, and R.C. Burton. 1995. OTTER: A testbed submersible for robotics research. Pp. 587–594 in Proceedings of the ANS 6th Topical Meeting on Robotics and Remote Systems held February 5–10, 1995 in Monterey, California. La Grange Park, Illinois: American Nuclear Society.

Webb, D. 1996. Personal communication to Donald W. Perkins, May 2, 1996.

Wiebe, P.H., C.H. Greene, T. Stanton, and J. Burczynski. 1990. Sound scattering by live zooplankton and micronekton: Empirical studies with a dual-beam acoustical system. Journal of the Acoustical Society of America 88:2346–2360.

Yoerger, D., and J. Slotine. 1987. Task resolved robust control of vehicle/manipulator systems. Pp. 17–26 in Undersea Teleoperators and Intelligent Autonomous Vehicles. N. Doelling, and E. Harding, eds. MIT Sea Grant Report 87-1. Cambridge, Massachusetts: MIT Sea Grant College Program.

Youngbluth, M.J. 1984. Manned submersibles and sophisticated instrumentation: Tools for oceanographic research. Pp. 335–344 in Proceedings of SUBTECH 1983 Symposium held November 15–17, 1983 in London, England. London: Society for Underwater Technology.

Yuh, J. 1990. Modelling and control of underwater robotic systems. IEEE Trans. on Systems, Man, and Cybernetics 20(6):1475–1483.

Zorpette, G. 1994. Autopilots of the deep. IEEE Spectrum 31(8):38–44.

3

Vital National Needs

Undersea vehicles have evolved into important tools for investigating the deep ocean and managing its resources. In their versatility and their power, these vehicles can be compared, perhaps, with satellites, although they operate in a vastly different realm. Like satellites, they expand the scope of human observation. A large fraction of the boundaries of the Earth's crustal plates are in the deep ocean and directly observable only by using undersea vehicles. Undersea vehicles give us access to vital information on our environment. Like satellites, they offer new perspectives and enable new tasks. They may even be the basis of entire new industries, such as subsea mining. The results of the first undersea explorations have fired scientists, engineers, and entrepreneurs with the universal desire to know more, to explore further, to extend the human reach. This chapter reviews some of the potential of those explorations for furthering the national interests in a range of scientific, regulatory, military, and industrial fields.

To illustrate these potential contributions, the committee has included the four focal projects shown in Table 3-1—technology development and integration projects that could lead to new applications of undersea vehicles in support of national objectives:

- synoptic observation system
- blue water oceanographic sections and hydrographic surveys
- subsea oil field inspection and intervention
- search and survey

The projects would make possible tasks heretofore unachievable or impractical but of obvious value to society. Some tasks rely on technology that is available and economically feasible today; others are more visionary. Each uses technologies of value to a wide range of undersea vehicle applications to meet commercial, military, and scientific needs. The committee has not ranked the projects as to priority. Many of these new primary applications would use existing technology for subsystems but would combine them in new and different ways. Some would be appropriate government missions, others would be led by the private sector. Certainly they do not exhaust the potential for innovative uses of undersea vehicles, nor are they intended as a representative sample of potential applications; they simply offer a few challenging examples.

All of the focal projects use AUVs as their primary vehicles. In the committee's judgment AUVs promise more payoff in advanced capabilities than DSVs or ROVs, which are technologically relatively mature. At the same time, DSVs and ROVs could be used in a variety of functions in these projects, not only in surveying and construction, but also as alternatives to AUVs in some missions.

SCIENTIFIC UNDERSTANDING AND APPLICATIONS

Many of the most important questions humankind can ask are problems of ocean science that require exploration of the little-known world beneath the surface: fundamental questions about the Earth's history; vital questions about the future of the human environment; and technical and economic questions. Nearly all marine activities require some scientific understanding of the ocean. Seagoing people have always charted the currents and depths and marked the migration seasons of edible fish. Over the centuries, these practical observations took on the character of true science. Modern oceanographic investigation is much more recent; the voyage of *Challenger* in the 1870s marked the first global-scale scientific cruise. The data gathered during its 3.5-year excursion filled dozens of volumes. Progress in oceanography has always been limited by access beneath the sea surface, and advances have depended on improvements in data-gathering tools and techniques. The emerging capabilities of undersea vehicles have made visible facts and relationships that are changing our central ideas about the oceans, their uses, and life on Earth.

TABLE 3-1 Focal Projects: Responses to National Needs

Focal Project	Opportunity	Vehicle	Key Development Concerns	Technology Maturity

Synoptic Observation System	Science
	  Sensing fine gradients over fixed areas, long term
	  Climate change
	Industry
	  Measuring ocean environmental change
		AUVs, with central "garage" station	Homing and docking; data transfer between subsystems	Component technologies available; no system experience

Blue Water Oceanographic Sections and Hydrographic Surveys	Science Sensing oceanographic gradients, extending range and accuracy of surface-supported surveys	AUVs	Positioning accuracy; integration of sensors, navigation, and control systems	Basic technology and subsystems available; systems integration needed
Subsea Oilfield Inspection and Intervention	Industry Reducing or eliminating dependence on surface support for subsea operations, maintenance, and inspection	AUVs	Integration of sensors, navigation, and control systems	Technology available; key components in development

Search and Survey	Science
	  Initial surveys for oceanographic research
	Industry
	  Search and recovery of high-value objects
	  Initial surveys for minerals
		AUVs	Batteries with higher energy density; control; data storage and processing	Technology available for most key components

Plate Tectonics

In the past several decades, undersea vehicles have contributed to profound discoveries in the Earth and life sciences that have emerged from human exploration of the seafloor. Science has described and verified processes of plate tectonics, consisting of continental drift and seafloor spreading resulting from the motion of the Earth's crust, which is driven by the slow, convective circulation of the molten interior. The distribution of continents—with their earthquakes, volcanoes, and mineral deposits—stems from these motions. The continental drift hypothesis was advanced in the last century by Alfred Wegener. This and the more recent concept of seafloor spreading have been confirmed by amassing and synthesizing large amounts of data from shipboard acoustic depth finders, marine gravimeters, magnetometers, deep sea dredges, and corers. In the 1980s, undersea vehicles gave researchers close-up views of hydrothermal vents, and cold seeps were discovered on the seafloor near tectonic ridges and faults, with their remarkable animal communities. These discoveries offer important fundamental insights into the functioning of the planet. They also have practical strategic and economic benefits, helping us determine the locations of important resources and protect ourselves from earthquakes and volcanic activity.

A central tenet of plate tectonics is that most geologic interactions occur in a small number of narrow, highly dynamic zones where the edges of large plates are in contact. There are three types of plate boundaries: deep ocean trenches, spreading centers or mid-ocean ridges, and transform faults. To understand key processes of our planet, including volcanism, earthquake patterns, and the formation of resource deposits, scientists must understand the processes occurring along these plate margins. With few exceptions, these active plate sutures lie beneath some portion of the ocean. A small percentage of these plate boundaries has been acoustically mapped from surface ships.[1] However, fundamental understanding of the processes along the three plate boundary types requires direct access, which can be accomplished only by undersea vehicles. Direct access is required for recovering geological cores, making in situ measurements of magnetics or temperature, taking close-up photos, and a host of other tasks.

---

[1] Acoustic mapping of the seafloor areas at and near seafloor active plate boundaries, including subduction zones, is estimated to cover on the order of 10 percent, or less, of the total area (Tucholke, 1995).

Deep Ocean Trenches

At the boundaries between oceanic plates and the continental plates, where tectonic motions compress the crustal materials together, the linked processes of lithospheric subduction and mountain building occur. It is along these lines of collision that the deepest places in the ocean are found. In the Pacific Ocean, this line forms an arc that runs from the tip of South America to the Aleutians, past Japan, and down to New Zealand, known as the "ring of fire" because of the prevalence there of volcanoes and earthquakes.

The deep ocean trenches are the least explored regions of the Earth. Even the Moon is better mapped and understood than these remote, hostile areas. Yet, because the trenches are the products of lithospheric subduction, one suspects there is much to be learned from probing them. Our experiences during the relatively recent investigations of oceanic spreading centers on the seafloor, with their hydrothermal vents and completely unsuspected thermophylic and chemosynthetic life forms, urges scientific exploration, surveying, and hypothesis-testing in the deep ocean trenches. Certainly that there is much to be learned. But the mechanical means for doing so are at present limited to towed sleds and a single Japanese ROV (*Kaiko*). Further development of undersea vehicles could extend the reach of these tools.

The deepest trenches—cold, dark, and under enormous pressures (1,100 atmospheres at the bottom of the 11-kilometer-deep Mariana Trench)—have only been penetrated twice, by Piccard and Walsh in the bathyscaphe *Trieste* in 1960 (Piccard and Dietz, 1960) and by the Japanese ROV *Kaiko* in 1995. Their observations at the bottom of the Mariana Trench show the presence of life, ocean currents near the bottom, and deep sediments of biological detritus from the water column above. The sediments in the trenches represent a loss of carbon-bearing biological materials from the near surface region; the sediments are an important sink for the removal of carbon dioxide from the atmosphere, as explained in the discussions on global warming and the carbon cycle later in this chapter. More observations will be made during later excursions of *Kaiko*, but that vehicle's limitation of mobility (owing to its long tether cable) will preclude any but the most preliminary exploration. Nevertheless, *Kaiko* and its companion, the DSV *Shinkai* (depth range 6,500 meters, with plans to increase this range), will be valuable assets in helping scientists identify new objectives that can be probed by more capable vessels carrying sensors, samplers, and instrumentation.

Scientific questions that come to mind span the entire expanse of ocean science. In geology and geophysics, they include: What is the heat flow-out of the bottom in these regions of high-friction lithospheric consumption? What are the levels and patterns of seismic activity? What gravity anomalies result as lighter crust is thrust downward beneath the continental margins, and what do they imply about the distribution of rock beneath the seafloor? What is the rock magnetism there, and how does it compare with the magnetism found closer to spreading centers? How rapidly are sediments accumulated from the debris of the overlying water column? Do the sediments slump into the trenches, partially filling them in (and thereby offering potentially attractive sites for seabed disposal of undesirable materials, via human-induced sediment slumps that would quickly cover over such materials)? In physical oceanography one asks: What are the currents and the temperature and salinity structures in the trench regions? Is thermally driven convection occurring in the water column? How much communication occurs with surrounding waters, especially the deep waters produced elsewhere by thermohaline processes? In chemistry, we wonder: What are the concentrations of trace materials in these areas of unusual geochemistry and hydrodynamics? In biology: What kinds of life forms exist there? Where do they derive their supplies of nutrients and energy? Are there opportunities for biotechnological developments? The answers to these questions are sure to be surprising and produce new targets for research.

A joint or international exploratory scientific program with the JAMSTEC, using *Kaiko* and *Shinkai*, could gather the information needed to formulate more precise scientific questions. In the meantime, preliminary design work for a deepest-diving vehicle capable of taking instruments and samplers directly to the seafloor could be performed. The design would have to be flexible and include ceramic pressure hulls and various in situ instruments (such as water samplers and temperature, salinity, and pressure sensors), remote sensors (acoustic profilers and mappers, video cameras, and laser scatterometers), geological and geophysical instrumentation (coring devices, small gravimeters, and magnetometers), and biological sampling devices. Although the initial design would likely be an AUV (with or without fiber-optic cable to the surface), ROVs and DSVs could be considered as well (Hawkes and Ballou, 1990). The preliminary design of the vessel and payload could lead to construction of a vehicle to help answer the many more detailed questions that would arise from the Japanese–U.S. program.

The profoundly different life forms found at the much shallower hydrothermal vents near tectonic ridges (discussed later in this chapter) allow the inference that the trenches also are likely to contain novel life forms uniquely adapted to their habitat. These organisms could be the source of new genetic material of interest to both science and industry. It is also conceivable that the trenches could be used for disposal of certain wastes. The combination of very deep sediment accumulations and earthquakes is likely to cause sediment slumps that would bury waste material, which would at the same time be slowly carried into the interior of the Earth by subduction. These possibilities are only conjectures because of the extreme scarcity of information about the trenches. While the trenches are the least explored regions of the Earth's surface, they are among the most active tectonically. If the lessons of discovery in other regions of the seafloor are any guide, important findings await discovery.

Undersea vehicles are essential in such exploration. Instruments deployed from surface platforms cannot be used; the length and weight of the cables to support such deep probes would require very large cable diameters merely to support their own weight, and enormous winches to handle 11 or 12 kilometers of heavy cable. One of the other platform options for exploration of the deepest depths, the human-occupied, dirigible-like *Trieste*, became obsolete and was retired in 1982. Now that the ROV *Kaiko* has reached the deepest ocean, the next step in the expansion of temporal and spatial observation and exploration would appear to be achievable using a deep-diving AUV or DSV equipped with ceramic pressure hulls, sophisticated in situ and remote sensors, and data acquisition and logging capabilities.

Spreading Centers or Midocean Ridges

A system of spreading centers stretches through most of the Earth's major oceans, like the seams on a baseball. Along these boundaries, the ocean floor spreads at rates of 20 km to 200 km per million years, creating new ocean crust. The new crust forms in water depths of about 2,000 to 4,500 meters as molten magma solidifies. In the porous ocean crust, high-temperature magma (1,200°C), meets near-freezing seawater. The water is heated, becomes buoyant, and carries dissolved minerals up through the crust, where they precipitate on contact with cold water at or near the seafloor. The mineral structures created by this process can form massive mineral deposits through which the hot hydrothermal fluid flows, exiting in dark plumes known as "black smokers." These venting sites and deposits were discovered and explored using a tow sled and an undersea vehicle during the first direct examination of a spreading center. The vents were located with the towsled *Angus*, and the *Angus* crew then directed the DSV *Alvin* to the site (Corliss et al., 1979). In the ensuing 16 years, underwater exploration, using a variety of vehicles, has located over 100 such deposits in all major ocean basins.

The heat and chemicals emitted by these vents sustain diverse chemosynthetic thermophilic microbial populations that form the base of a widely distributed and previously unsuspected food chain that is not based on photosynthesis. Studies of the chemical and energy flows and feedback loops in submarine volcanic systems also raise central concerns of planetary science because conditions at the vents are analogous to the oxygen-poor conditions on the early Earth and on other volcanically active planets. Characterizing them will require mapping the vents and sampling their emissions at a variety of scales, using multisensor survey tools capable of precise, high-resolution mapping. Measuring the variations of the vents with time requires repeatedly deploying, servicing, and recovering instruments in corrosive and hostile environments. Performance of many of these functions will require undersea vehicles.

Recent use of a U.S. Navy sound surveillance system to identify and locate seismic activity associated with diking and eruptive activity on the Juan de Fuca Ridge allowed rapid scientific response to study such an event for the first time ever (Fox, 1995). The ability to respond rapidly (within days) to such events using undersea vehicles will be increasingly attractive to scientists.

Transform Faults

In transform faults, such as the San Andreas Fault, plates slide past one another, without gaining or losing mass, occasionally generating large earthquakes. Undersea vehicles have been used to examine transform faults (Fox and Gallo, 1984). Because the faults are usually deep, steep-walled valleys oriented at high angles to the trends of spreading centers and subduction zones, they offer a rare cutaway view of the adjacent crust. Thorough mapping of the walls of such a fault using either extensive DSV observation time or side-scan sonar sheds light on the stratigraphic record of the volcanic crust adjacent to a ridge crest. Follow-up studies with vehicle-mounted drilling tools could allow major strides in characterizing this geologic history.

Essential vehicle requirements for investigations in transform faults include high-precision mapping in steeply sloping, extremely rugged terrain; precision navigation and platform control on preexisting maps; and innovative sampling strategies involving horizontal drilling techniques applied to near vertical walls.

Geochemical Cycles

One great challenge facing oceanographers over the next few decades will be predicting the ocean's responses to natural and human-induced environmental change, such as climate change and land use changes that affect the coastal oceans. Because the deposition and transport of chemicals in and by the sea affect such processes, observations of the fate and effects of chemical species in the ocean can provide sensitive indicators of global variability. Of special importance is the role of the ocean in the global carbon cycle because of the uptake and release of organic carbon by the sea. As atmospheric carbon dioxide concentrations rise and the potential for climatic warming increases, it becomes essential to know if the oceans sequester or release carbon dioxide and under what conditions they might do either. Vast quantities of carbon are stored in the sea in a variety of forms in sediments and in totally dissolved inorganic carbon. Marine organisms transfer carbon to the ocean bottom. The cycling of carbon in the oceans, and the extent to which the oceans are a net sink for the increased atmospheric carbon dioxide, are far from clear. The role of hydrothermal vents in the carbon cycle is understood only in part.

In some research areas, such as the study of fluids emitted from hydrothermal vents and cold seeps (including oil seeps), DSVs are essential tools (Edmond and Von Damm, 1979; Von Damm et al., 1985; Guinasso et al., 1994). For chemical oceanographers, the primary advantage of DSVs has been their ability to obtain samples from very precisely chosen locations (e.g., inside a vent orifice or in the nepheloid layers sometimes called "fluff" sediments). Other investigators have also taken advantage of the ability of DSVs to carry out complicated manipulative tasks on the seafloor. Using ROVs or AUVs for such missions would require the development of better depth perception to positioning samplers and very fine navigational and positional control, but it might be worthwhile for reasons of cost, safety, and operational efficiency. In addition, payload capacity will have to be sufficient for a variety of tasks, including obtaining several samples (on the order of 1 liter each) from small vents without entrainment of bottom seawater and deploying, manipulating, and retrieving equipment packages from the bottom. Among the most demanding sampling tasks are sediment coring and the collection of organisms from hydrothermal vents.

In recent years, a variety of vehicles, including benthic landers, have been developed to permit remote operation on the seafloor. Although most of these benthic landers drop to the seabed, land, and collect sediment or remain on the bottom while intermittently sensing or sampling, then return to the surface, they are not self-propelled vehicles according to the definition applied by the committee. However, *Rover*, a bottom-transecting vehicle device developed by the Scripps Institution of Oceanography, is designed to land on the seafloor, intermittently activate itself, drive to a new spot, collect, sample, and return to the surface on command after deployment as long as a year (Smith, 1995). ROVs have been used to map chemical distributions near a vent (Coale et al., 1991), and an ROV has even been used to sample from a vent. These vehicle applications are indicative of the potential of more advanced vehicle systems. To illustrate the potential of AUVs in delineating the fine structure of variations in the ocean waters over time, the committee has developed a focal project on a synoptic observation system, described in Box 3-1.

ROVs and AUVs could improve survey coverage in programs that require monitoring chemical species (for example, carbon dioxide) in areas where weather or ice conditions do not permit the use of surface ships or moorings.

One factor limiting the use of undersea vehicles in chemical oceanography has been the lack of chemical sensors rugged enough for extended deployment. Among the sensors already available are chemical sensor systems, such as electrodes or fiber-optic probes for oxygen or pH, and scanners (submersible chemical analyzers) for nutrients (metals and hydrogen sulfide). Many new and improved sensors capable of long-term deployment are being developed, including sensors for carbon dioxide, oxygen, pH, and other chemicals. The National Research Council recently published a report addressing chemical sensors for ocean carbon-cycle studies (NRC, 1993). It emphasizes the importance of in situ sensors, calibration, and quality control. It also sets priorities for target chemicals and sensor development in the following order:

- quantifying anthropogenic carbon input (various carbon and carbon dioxide system measurements
- understanding the biological pump (measuring organic compounds)
- tracing water masses (CFCs and other tracers)
- other analyses (Zn, Cu, Al, Pb, Mg and other elements of interest to chemical oceanographers)

Another high priority is the development of sensors for measuring nutrients. The most useful sensors will be those that can be attached as needed to many types of vehicles.

Dynamics of the Ocean, Atmosphere, and Sea Ice

The ocean plays an exceedingly important role in establishing weather and climate over land and sea alike. In turn, it is profoundly affected by the overlying atmosphere and ice cover, while serving as the flywheel on the coupled system. Within these conjoined geophysical fluids, life has evolved and flourished, with the very habitability of the Earth depending on the sea to an extent not very well appreciated. The tropical oceans are the dominant absorbers of the sun's energy; through complex processes involving evaporation, cloud formation, initiation of winds, rainfall, and the redistribution of solar heat by winds and currents, the air and sea together carry heat, momentum, and moisture from the tropics toward the temperate and polar regions. In the process, they make life as we know it possible.

Understanding these overarching global processes is the business of physical oceanography and atmospheric physics. The general circulation of the ocean and atmosphere is a complex phenomenon spanning the planet and is one that requires diverse measurements over large and small scales alike. But measurement alone is not enough. The elucidation of the many physical, chemical, and biological processes involved requires an understanding of the mechanisms at work, their successful theoretical descriptions, and their incorporation into large-scale numerical models that are capable of predicting the state of the air and sea. This is the business of weather and ocean forecasting.

Because of the enormous extent of the sea, oceanography has always been a data-starved science, and marine scientists have resorted to a variety of means to garner information. The classic approach is to use sensors and samplers lowered from a ship to measure at depths such properties as temperature, salinity, pressure, and chemical substances along routes at sea termed "oceanographic sections." More recently, indirect sensors, such as fluorometers and transmissometers, have been added to the research ship's suite of subsurface instrumentation. Other techniques include mooring current meters, thermistors, pressure gauges, and other

BOX 3-1
Focal Project 1: Synoptic Observation System

Opportunity. Advances in understanding and modeling submarine volcanic or oceanic processes have brought requirements for fine-scale temporal and spatial measurements of various parameters in large, defined regions. Modeling also often requires knowledge of higher-order derivatives of parameter gradients, which places further demands on the resolution and frequency of measurements. Current capabilities for sampling broad areas synoptically are largely nonexistent, but those capabilities would be of significant use to industrial, scientific, and even military interests in measuring and characterizing specific ocean and seafloor volumes. Sensors for making these measurements have improved to the point that, for most applications, they are no longer a major limitation; rather, the restricting element is the platform necessary for sensors or sensor data transport.

Objective. Provide a method of sensing synoptically fine-scale gradients in the water column or seafloor over wide areas (several hundred km^2) and long term (up to several years), and record and transmit these data via ship or satellite relays to the scientist or commercial user without full-time dependence on expensive surface support ship(s) and services. Provide quick-response data acquisition in areas likely to experience rapid changes in geological or biological character. Developments in AUVs and complementary technologies have evolved to the point that the construction of long-term "undersea observatories" is possible.

Vehicle System. An "undersea observatory" for synoptic measurements is illustrated below. Such an observatory would consist of a central control base, with a tether mooring to a surface communications buoy and facilities to "garage" one or more AUVs, and several outlying "nodes," each with an array of instrumentation, a record and playback capability, and a cooperative "dock" for mating with the AUV.

Either on command or by preset timing, an AUV disconnects from its "garage" at the control base and proceeds to the first preinstalled sensor node, navigating on dead-reckoning until it approaches within the "homing" sector, wherein the AUV repeatedly interrogates the node and elicits vectoring acoustic responses. As the AUV approaches to within fine-scale range, the node switches to either optical or high-frequency acoustics to guide the vehicle into a docking "socket." Once the vehicle and socket are joined, the node dumps its prerecorded, time-based data into the vehicle on-board recorder through an inductive coupler or optical window and resets its own recorder. The AUV then resets its dead-reckoning navigation system, disengages, and proceeds to the next node, repeating the process until all nodes have been served.

After storing all node data, the vehicle returns to the control base, homing and docking in its "garage," where it dumps the stored data to the control base, which sends the information to a land site via a surface buoy and commercial satellite communication such as INMARSAT. Received data can then be directed to specific sites or put on the Internet for broader distribution. During the interval between repeat circuits to the outlying nodes, the vehicle's batteries are recharged, and the onboard computers receive command updates from the land site through the satellite link, the surface buoy, and the communications connection between the garage and the vehicle. In addition to servicing nodal regions, the AUV could carry selected sensors, either providing or augmenting time-based data during transits or, upon command from shore, direct itself to specific areas for rapid reaction surveys of seafloor events detected by the control base instruments or other means. When such an event occurs, a human may assume task-level control of the AUV in near real time to take immediate advantage of its presence at the scene. Later, as desired, the AUV can be released to resume its preprogrammed rounds. Ongoing advances in acoustic communication will make possible low-cost monitoring of the AUV's location and functions, allowing for quick recovery from failures of autonomous systems.

Fiber-optic cables could be used for linking the nodes, with AUVs deployed for time-critical measurements. However, the specified system is more robust and better suited for severe environments such as spreading centers and other seismically active areas. In addition, the technology of housing and docking is important in itself. Use of acoustically linked nodes is another option for some applications covering small areas. Acoustic communications offer greater flexibility for system changes and probably lower cost installation than fiber optics. Another option is the use of AUVs for sensing between nodes to resolve issues of spatial heterogeneity.

This project would also lend itself to the use of multiple low-cost AUVs dispatched according to signals from the field of sensors to make time-critical measurements.

Performance Requirements:
- Area coverage: 900 km^2
- Range: 400 km
- Depth: 3,000 m
- Speed: 3 kt (average), 5 kt (maximum)
- Length: 8 m
- Diameter: 1 m
- Navigation: Dead-reckoning using Doppler sonar and inertial navigation system
- Obstacle avoidance sonar
- Nodal acquisition: Dual-stage, initial long-range acquisition (2 km), use low-frequency sonar; fine stage acquisition (< 50 m), use high-frequency sonar or pulsed light beam
- On-board sensors: Chemical, salinity, temperature, depth

Analysis of State of Technology and Practice. An underwater observatory system would require a number of technologies to act in concert over long periods of time. Many of these technologies have been demonstrated in restricted engineering environments, but a representative network has not yet been assembled and used. The major development concerns for this focal project are the nodal stations—the vehicle homing, docking, and data dumping—and the "home garage" with its surface communications.

Nodal station subsystem — This subsystem serves as the primary collection and recording point for all sensors in the surrounding area. The station has to be reliable for long periods of time, perhaps extending to years. The system used for cooperative homing and docking with the vehicle must function flawlessly.

Subsystem for vehicle homing, docking, and data pass-through from the node — This subsystem will use new technology that has little or no history. Subsystem components, including optical windows and connectors, are available with some adaptation needed, but significant software development will be required.

Garaging subsystem — This subsystem is a major systems integration and reliability challenge. In addition to providing a homing and docking capability similar to that for the data collection nodes, the garage must provide reliable battery-charging, data transfer, and storage capability and a communications component that provides for transfer of collected data to the surface through either acoustic telemetry or a buoy system. Seawater batteries could be used where environmental conditions are favorable (as in the Labrador Sea and many other areas of scientific interest). These batteries are commercially available, and the size would not be a handicap if they were used to power the stationary charging system.

Synoptic Observation System

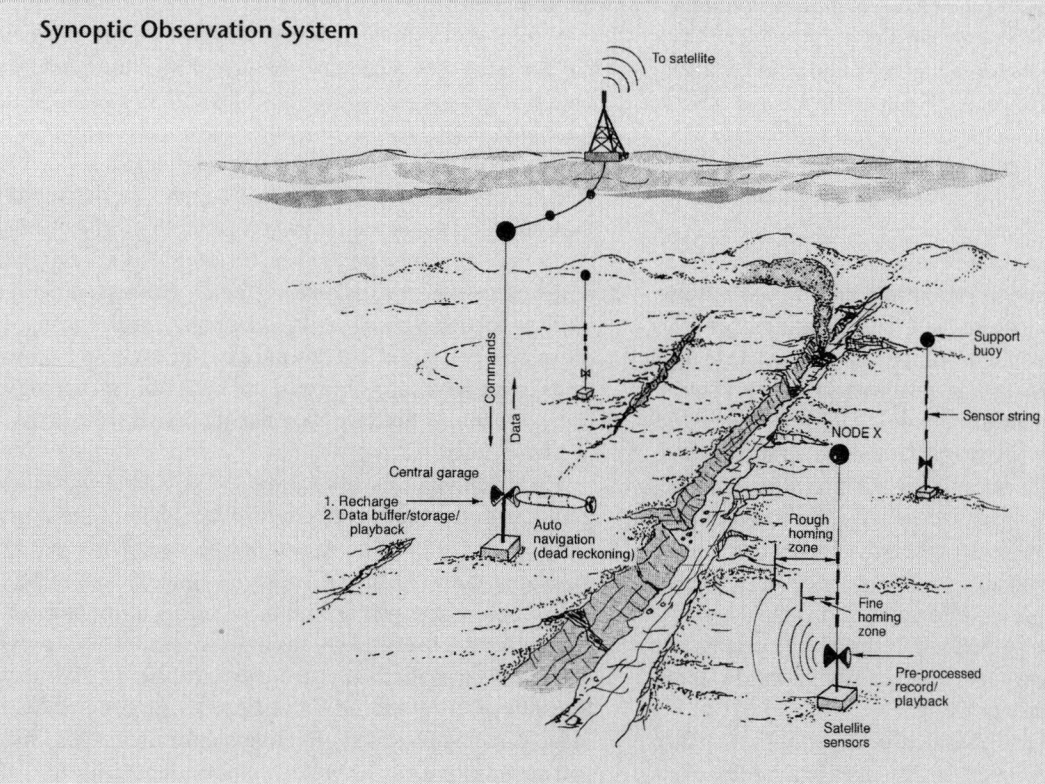

instruments to cables anchored on the seafloor to acquire time series of current velocity, temperature, and the like. Drifting buoys, both surface and subsurface, carry instruments and positioning devices that allow them to sample properties while acting as passive tracers for water circulation. In the last two decades, satellite remote sensing has grown to prominence, and its methods give near-global measurements of limited but important surface quantities such as temperature, chlorophyll and sediment concentrations, currents, wave heights, wavelengths, and wind velocity. However, satellite measurements are generally confined to the surface or near-surface of the three-dimensional volume of the sea and therefore can tell little about events going on beneath it. Undersea vehicles can obtain novel, extended, or otherwise unobtainable data on current, density, chemical, or biological composition and other ocean properties. For example, currents may be measured in strong flow at depths beyond the reach of shipborne acoustic Doppler current meters; wave spectrum measurements may be made under storm conditions (when buoys lose their measurement capacity) by sonar observing upward from beneath the waves; and horizontal measurements of many properties may be made simultaneously from the ship and from an AUV running a parallel course at some distance from each other, thereby determining these gradients at constant time.

Examples of using undersea vehicles in support of physical oceanography are numerous. One intriguing application is to use AUVs under the sea ice canopy to gather data otherwise exceedingly difficult to obtain. Since air-sea interchange in the Arctic proceeds through leads and polynyas (openings in the ice), the Arctic Ocean is in poor communication with the atmosphere during the polar winter. Very little is known about polar oceanography under these conditions. The ability to send an AUV under the ice and have it return such important information is a clear application of an undersea vehicle (Curtin et al., 1993). Another application is measuring storm waves by upward-looking sonar on an AUV. Wave-prediction models are exceedingly difficult to calibrate for storm conditions because many of the observations needed for calibration and verification would put humans at risk. In places where there are surface buoys with appropriate sensors (as in the Gulf of Mexico), reliable data can be obtained. However, for most sectors of the oceans, sending AUVs into a hurricane would yield new data and insight on surface wave spectra, bubble generation, wave plunging, and upper-ocean mixing processes under high wind conditions. AUV sonar and imaging devices can provide unprecedented data under otherwise inoperable conditions. Such data are valuable to designers of ships or oil-drilling rigs, for example. A third application is to use semi-autonomous vehicles that drift with subsurface currents but periodically adjust their depth and position to follow some desired criterion of water property.

While the focal project in physical oceanography (see Box 3-2) is perhaps less dramatic than the ones just discussed, it will be of considerable value in terms of both science and economics. Blue water oceanographic sections and hydrographic surveys using undersea vehicles show how the efficiency of such data acquisition from an oceanographic research ship could be doubled or tripled at only incremental costs. In addition, and perhaps more important, using an AUV in conjunction with a surface ship would provide information on gradients in the sea that is not otherwise attainable. Since gradients drive both the hydrodynamics and thermodynamics of the ocean, the AUV would provide increased knowledge of some of the fundamental forces that activate the ocean. Using AUVs would also greatly increase the efficiency of hydrographic surveys. While there is little experience to date in the use of submersibles to support physical oceanography and surveys, the potential is clear and exciting.

Productivity of the Oceans

Productivity is an energy-based measure of the ocean's biological processes. At present, oceanic productivity is best understood at the base of the food chain, where plants fix inorganic carbon into organic compounds using solar energy. Recently scientists have learned that symbiotic bacteria living in the tissues of hydrothermal vent animals are also primary producers, using chemosynthesis rather than photosynthesis. Further up the food chain, scientific understanding decreases. This situation is in part the result of increasing complexity at the higher levels of organization, but it is also the result of limitations of the technology available to study these organisms.

While scientists have been studying oceanic communities for more than a century, the perspective afforded by research vessels at the surface has biased our knowledge substantially. For example, scientists have underestimated the pelagic biomass by as much as one-quarter (Robison, 1995; UNOLS, 1994) because gelatinous animals are destroyed by the collecting nets used to bring them to the surface and because they are invisible to acoustic scans. Recent estimates, based on surveys by undersea vehicles, are beginning to revise our understanding and perceptions about oceanic life processes, but this effort has only begun. Nets and acoustic assessment have also obscured the behavior and ecology of oceanic animals because they cannot reveal their activities in useful detail.

Undersea vehicles have provided scientists and investigators with the tools to begin to solve some of these problems by giving them an in situ perspective. While net tows measure patterns at scales from ten to hundreds of meters in the vertical plane and at kilometer scales in the horizontal plane, direct observations from ROVs and DSVs provide resolution on spatial scales from a centimeter to a kilometer. Recently, ROVs and DSVs equipped with high-resolution video cameras provided effective platforms for this kind of marine exploration in water column depths up to 1,000 meters (Hamner and Robison, 1992; Matsumoto and Harbison, 1993). These techniques have also been applied to comparable studies of bioluminescence (Widder et al., 1989) and marine snow (Pilskaln et al., 1991). With the extension of these studies into deeper water, a large number of undescribed species and "new" ecological relationships are coming to light. Among the ecological features revealed by these measurements are animal associations, habitat partitioning, the role of substrate in spatial distribution patterns, and animal densities. Undersea vehicles allow observation of trophic relationships, physiological rates and processes, activity levels, behavioral patterns, standing stocks, and reproductive patterns (Bailey et al., 1994; Robison, 1995; Adams et al., 1995a). Undersea vehicles can play a major role in forthcoming studies of marine biodiversity (NRC, 1995).

The need to understand the dynamics of marine animal populations and their coupling and the physical and biological processes in their environment is fundamental to the larger issue of how global climate change affects marine systems. One of the U.S. components of the Global Ocean Ecosystem Dynamics study, the Georges Bank program, has focused on studying the population dynamics of four target species of plankton on the Bank (historically one of the world's great fisheries production areas). However, cod and haddock populations have been driven so low by overfishing that these fisheries on Georges Bank are now closed. Integrated surveys of distribution and abundance of the animals, process studies of the predators and prey, and the impact of major physical processes on the biological rates are required. The use of ROVs and AUVs, while limited in the initial stages of this research, holds promise for enhancing the understanding of how individual species exist in their environments on very small scales and the time-series changes that take place over periods of months or years.[2]

On the floor of the ocean, science is making similar progress. In the course of its long career, the DSV *Alvin* has made a number of fundamental discoveries that significantly altered concepts about life on the deep seabed. After the discovery and exploration of hydrothermal communities by a team operating a tow sled and the DSV *Alvin* in 1979, subsequent diving programs found vents in several other locations. Even more widespread are the various cold-seep communities, with fauna related to the vent animals but driven by chemical compounds such as methane and hydrogen sulfide. These discoveries, all made and sustained by undersea vehicles, have fundamentally changed scientific understanding of the basic relationships governing life on Earth.

Biotechnology and Biomedical Marine Research

The ocean's animal and plant species form an enormous genetic resource for producing a variety of chemical compounds that can be used for medical, agricultural, and industrial purposes. It is likely that many underseas genetic substances will be discovered and assessed; some have already been developed and have shown potential. The antimicrobial steroid, squalamine, recently isolated from sharks, appears to be a potent deterrent, with unusually slight side-effects, to fungal infections of the type that plague cancer and AIDS patients (Stone, 1993). Many other marine organisms produce similar antibacterial, antiviral, and antifungal agents.

Particularly promising sites for the discovery of new bioactive molecules can be found at the hydrothermal vents that occur at great depths along the ocean ridges. The biological communities there contain enzymes and other compounds that can function at very high temperatures and may provide thermally stable compounds for use in high-temperature catalysts, cleaning agents, or solvents. Deep-sea organisms are also sources of pressure-tolerant compounds. One value of pressure-tolerant catalysts, for example, is that their structural hardiness allows them to function effectively in challenging industrial applications (Flam, 1994; Jannasch, 1995; New England Biolabs Catalog, 1994; Adams et al., 1995a).

Undersea vehicles have already performed much of the work on hydrothermal vents, and both DSVs and ROVs are at work in the Pacific and Atlantic oceans, locating and collecting specimens for research. Undersea vehicles must be able to perform deep water operations to locate and survey sites of interest and then inspect them in detail over relatively small areas.

LIVING RESOURCES AND ENVIRONMENTAL MANAGEMENT

Living Resources

Management of marine living resources has grown more challenging as human populations have grown and fishing technology has advanced. Society needs more accurate ways to predict the sustainable yields of commercial fish stocks, including both traditional and nontraditional species. Problems in several of the world's fisheries demand better fishery management information. Such information must rely on data about population variables such as rates of birth, growth, and death; predator-prey relationships; spawning-ground requirements; nursery ground characteristics; the impacts of intensive fishing, and environmental controls. Traditional methods of collecting these data have reached their limits. Undersea vehicles can directly collect some of this information to characterize the state of communities in fishing areas. More important, they can provide new kinds of data (such as behavior patterns, predator-prey intersections, activity levels, and escape responses) critical to understanding these processes (Krieger, 1993; Adams et al., 1995b; Krieger and Sigler, 1996).

---

[2]AUVs would require video technology such as that described by Davis et al., 1992.

BOX 3-2
Focal Project 2: Blue Water Oceanographic Sections and Hydrographic Surveys

Opportunity. In oceanographic and hydrographic survey work, the operation of one or two AUVs in synchrony with a surface ship would provide a way around the severe limitations imposed by a single ship making measurements along a track. Coordinated data acquisition by two or three platforms would extend horizontally the vertical profiles made by lone vessels, with little additional personnel or operating cost. The data acquisition rate of the ship would be effectively doubled or tripled (or, alternatively, the vessel's hourly costs reduced by one-half to two-thirds), while the quality and simultaneity of the data would be increased. Among the projects that would benefit from such operations are (a) the taking of physical oceanography "sections" composed of sequences of "stations" along a path, usually hundreds to thousands of kilometers long; and (b) the conduct of hydrographic depth surveys or depth profiles made along similar sections for the purpose of charting the seafloor.

Objective. Large-scale currents are driven by surface wind stress and density gradients. The currents profoundly condition weather and climate. Reliable three-dimensional models of mesoscale phenomena, such as eddies and convection plumes in the Gulf Stream or cold water plunging in the Labrador Sea, require high-density parametric measurements not available from moored instruments or satellites. Today, establishing the current velocity and transport of water mass, momentum, heat, and salt of a current system requires transitting a research vessel (indicated by "RV" in the accompanying figure below) approximately perpendicular to the estimated direction of the surface flow. Moving at 6 to 12 knots, the vessel acquires data from sensors such as acoustic Doppler current profilers (ADCPs) and other towed instruments such as temperature and salinity sensors. At intervals of perhaps 20 to 50 km the ship stops and takes a "cast," lowering instruments via cable to measure conductivity, temperature, and depth with a CTD profiler and water mass properties, including chlorophyll and dissolved oxygen concentrations, sound speed, light transmission, and water samples. Without an AUV, the RV might have to run a reverse course, perhaps 10 to 50 km removed from the first, to sample spatial variability. During the time it takes to run the return path, however, the ocean has changed, so this method mixes spatial differences with temporal differences (a classic problem in observing fluids). It is the gradients in fluid properties that are important in the hydrodynamics and thermodynamics of the ocean and atmosphere, and it is important to establish the gradients through measurements that are spatially separated but simultaneous, as well as making point measurements that extend over time at fixed locations. The latter are usually accomplished via instrumented moorings. Thus, the use of two (or three) platforms with similar instruments traversing parallel paths allows the simultaneous measurement of spatial gradients in addition to doubling or tripling the area covered.

In acoustic depth-sounding, time changes do not arise, so the benefits of simultaneous operations are mainly in efficiency and time savings. However, precision depth sounding requires observations of upper water column temperature measurements to correct for the speed of sound in seawater; so vertical casts are required here as well.

Vehicle System. The figure below illustrates the operational concepts in oceanographic section-taking. The surface research vessel transits while using ADCPs and towed instruments. During transits, the AUVs execute sawtooth or square-wave dive profiles down to perhaps 1,000-meter depth, depending on the quantities to be measured, while running at the same speed as the RV and parallel to it, perhaps 20 km away. At each surfacing, the AUVs acquire GPS fixes to calibrate the inertial navigation systems, used while underwater. Radio frequency communications with the ship may also be needed. The AUVs also use ADCPs while under way and make continuous measurements of temperature, conductivity, depth, sound speed, dissolved oxygen, chlorophyll concentrations, et al. The RV and the AUVs then stop on their stations, the surface vessel to take a section and the AUVs to take a navigation fix, and transmit selected sample data via radio link to the ship or a satellite. After a distance of nearly 1,900 km (1,000 nautical miles), the AUVs are recovered, additional data is downloaded from their solid-state memories, new batteries are installed, and they are relaunched. This process should take only a few hours. The measurement procedures are repeated on each transit/on-station cycle, allowing a three-dimensional view of the ocean volume on three vertical planes (containing the RV data and the AUV profiles). As in Focal Project 1, a human may assume task-level control of the AUV whenever it is advantageous to do so.

In the case of acoustic depth-sounding, the AUVs would probably be closer to the survey vessel (perhaps 5 km away) and would remain near the surface but out of the wave zone for stability. The navigation requirements are somewhat more stringent, requiring more frequent GPS fixes, perhaps via a trailing wire. This arrangement would triple the swath width of a ship's coverage. The savings would be considerable; survey vessels cost $15,000 to $20,000 per day to operate.

Performance Requirements:
- Range to 1,500 km
- Depth to 1,000 m
- Speed: 6 to 8 kt
- Dimensions: 6 to 7 m long, 1-1.2 m diameter
- Mission duration to 160 h
- Working volume 4.0–4.5 m^3
- Launch and recovery from host vessel via A-frame or ramp
- Control by programmable tracklines, adjustable via radio frequency link to ship upon surfacing
- Navigation and position by GPS while surfaced, digital compass, inertial navigation at high latitudes, obstacle avoidance sonar for benthic missions
- Communications, radio frequency telemetry to ship or satellite, radio frequency homing link
- Sensors, ADCP, CTD, dissolved oxygen, chlorophyll concentration, sound speed, surface profiler, light transmissometer
- Work systems/tools, water sampling with multiple containers
- Processing
  — for vehicle, sufficient to support mission logic
  — for data, field programmable, solid-state memory, magnetic storage, fusion of position and sensor data streams

Analysis of the State of Technology and Practice. The limiting technology is in seawater batteries, but Norwegian aluminum-seawater batteries have demonstrated the required capability. Development and adaptive efforts (but not new technology) are needed for low-power-drain inertial navigation systems; radio frequency antennas that can function when wet with seawater; sensors with low power drain and small volume; and integrated sensors, navigation, and control systems.

Oceanographic Survey AUV in Coordination with Research Vessel

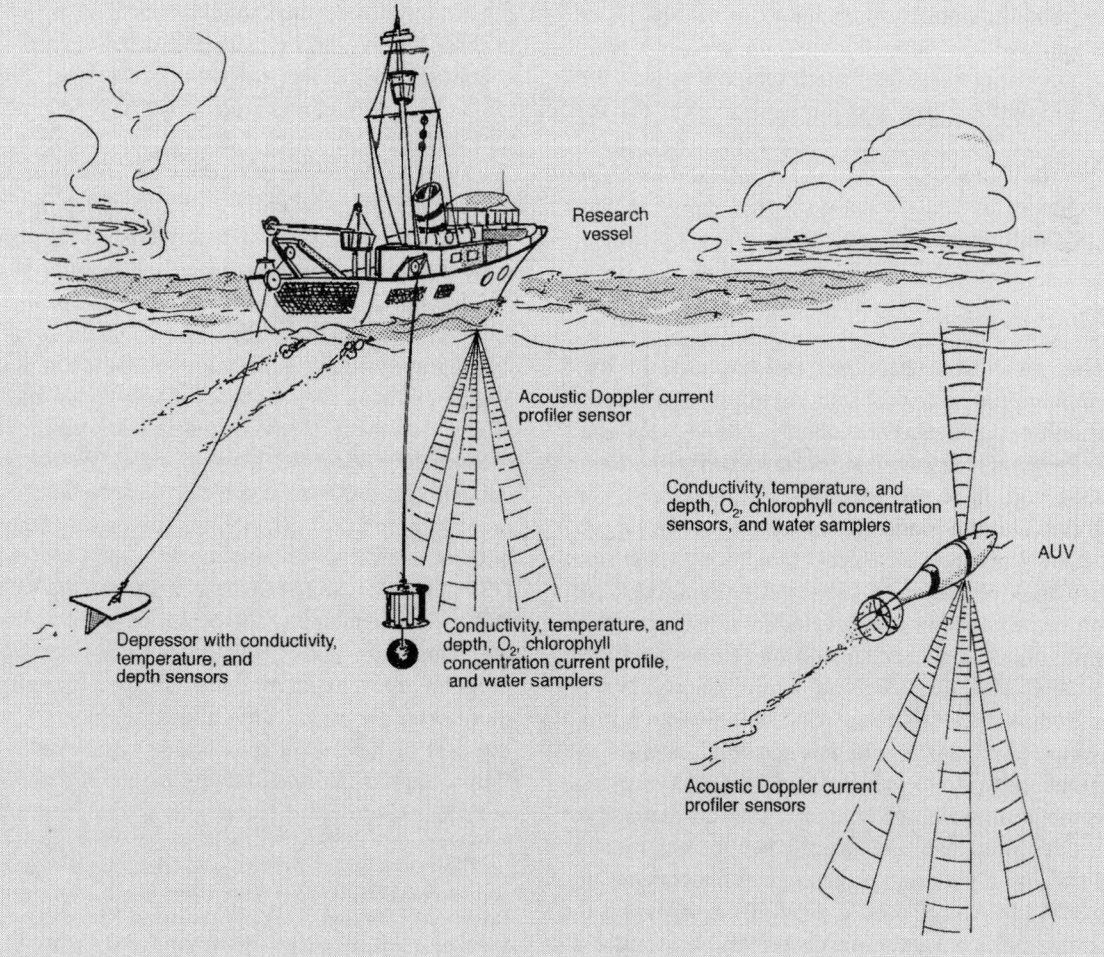

Undersea vehicles may be the only practical method of conducting detailed research on deep-water fisheries. Such missions would require large-area surveys in the water column, on the order of hundreds or thousands of kilometers. To provide reliable and repeatable results, these missions would require accurate navigation. Most importantly, fishery stock-assessment vehicles would have to use specialized sensors to perform the actual stock assessments (i.e., determining the size and type of the stocks under study). For research purposes, these surveys would be relatively infrequent (months to years) but fairly intensive during particular projects.

NOAA's National Marine Fisheries Service surveys 325 stations annually to assess fish stocks in U.S. waters from the Canadian border to Cape Hatteras. Each station consists of a 30-minute bottom trawl at about 3.5 knots. These survey tows represent an annual sampling of less than 0.01 percent of the area (U.S. Northeast Atlantic waters) for which stocks are being assessed. Even when the resulting data are supplemented with information on fishing effort and landings, this level of sampling leaves considerable uncertainty about actual fish stocks (NOAA, 1993). It is possible that AUVs with appropriate high-frequency acoustical sensors, integrated with laser-based optical/video systems, could provide better sampling coverage and better resolution in a cost-effective way. Such vehicles would also enable real-time studies of the effects of trawls on the seafloor and on benthic feeding processes. Operating costs for fisheries survey vessels are $5,000 to $15,000 per day, and the annual effort for the northeast region of the United States is about 320 days at sea per year. For the United States, overall annual effort for these surveys involves 1,860 days at sea at a cost of approximately $9.3 to $18.6 million per year.

Marine Conservation

Undersea vehicles are being increasingly used in programs to monitor the ecological status of marine sanctuaries and other subsea habitats. For example, NOAA's Marine Sanctuary Program is responsible for stewardship of a large and growing portion of this country's coastal waters. The Monterey Bay and the Florida Keys are two of the largest and best known marine sanctuaries. Monitoring these and other sanctuaries constitutes a large and growing task for NOAA. In some cases, undersea vehicles already perform habitat assessment and characterization (Robison, 1993; 1995). In the Florida Keys National Marine Sanctuary, researchers from NOAA's National Undersea Research Program have used submersibles to date sea-level changes by sampling reefs at depths to 140 meters and for verifying data sensed remotely from ships. NOAA also used a submersible to observe, describe, map, and quantify benthic community structure and its relationship to geological phenomena. In addition, ROVs have been used in depths from 40 meters to 100 meters to conduct surveys of deep reefs and to assess and characterize habitats. Future requirements will include long-term monitoring and regulation as well as rapid responses to serious perturbations. Undersea vehicles can perform these activities, and NOAA's National Marine Sanctuary Program is integrating these new tools into its regulatory work.

Waste in the Ocean

Wastes in the ocean come from many sources, including runoff from land, legal and illegal ocean dumping, and accidents. Debate continues about the ocean's appropriate role for disposal of other substances, such as sewage sludge, radioactive waste, and municipal wastes (Calmet and Brewers, 1991). The extent and disposition of the wastes in the ocean is rather poorly known. Responsible management of wastes depends on better knowledge of their natures and impacts. Existing sites may require long-term monitoring and, in some cases, perhaps, remediation. Prospective sites need thorough assessment, and undersea vehicles can play an important role in these activities.

The London Dumping Convention[3] sets terms for ocean dumping, but the list of legal substances has been steadily reduced in recent decades; most legal dumping in the United States is of dredged material and drilling muds. Illegal dumping has introduced substantial amounts of wastes, including at least 16 nuclear submarine and ice breaker reactors (6 containing radioactive fuel) dumped by the former Soviet Union in arctic marine waters (OTA, 1995). Most coastal nations have discharged industrial and human wastes. Accidents have introduced hazardous cargoes, nuclear reactors, and nuclear weapons into the oceans with poorly known results. Even well-known sites, established legally and deliberately, have generally not been well monitored.

Monitoring requirements vary from site to site. However, they generally include various combinations of studies in hydrography and circulation, radiochemical speciation, source and sink terms, ice dynamics, river and watershed effects, sediment dynamics, impacts on benthic fauna, and food-chain transfer (McDowell, 1994). When sampling and collecting at these sites, vehicles must be able to shut down propulsion to avoid disturbing sediments and must have the payload capacity to carry samples to the surface. In general, DSVs are better suited to these tasks than ROVs and AUVs. Better understanding of the deterioration of containers, such as drums and reactors, is also important because any new programs for legal dumping would generate substantial requirements for monitoring and assessment. The proposed disposal of high-level radioactive wastes in the deep seafloor, if approved, would likely require the use of undersea vehicles, given the great depths at which it would occur

---

[3]The London Dumping Convention is the general title given to the Convention on the Prevention of Marine Pollution by Dumping of Wastes and Other Matter, adopted in 1972 by the United Nation's Intergovernmental Maritime Organization (now the International Maritime Organization).

(> 4,500 meters) (Calmet and Brewers, 1991). Vehicles would survey and sample sites prior to emplacement, verify emplacement, and perform long-term monitoring.

Site 106, the deep-water sewage sludge dump site off the New Jersey coast, is an example of thorough monitoring and a good illustration of the utility of undersea vehicles in the monitoring process (Van Dover et al., 1992; Hill et al., 1993; Bothner et al., 1994). Barges have dumped inventoried sludge at specified positions and prescribed rates. Multidisciplinary teams studied dispersion, transport, and remobilization of contaminants and their effects on biomass, diversity, and organism body burdens of toxic metals and organics. Underwater vehicles, both DSVs and ROVs, played central roles in these studies, recovering samples, taking measurements, performing surveys, and recording observations (Bothner et al., 1994). An AUV may participate in this program in the near future as well. These studies have had important findings, but Site 106 is unusually well characterized.

The Naval Research Laboratory, working with industry and academic researchers, has begun a $1.5 million program to assess the environmental, technical, and economic aspects of isolating industrial wastes, such as sewage sludge and dredge spoil, on the ocean's abyssal plains, more than 3,000 meters deep. These areas are believed to be geologically stable and to experience only slight water currents. The Naval Research Laboratory will assess several specific sites and prepare a site survey plan, monitoring program, and economic analysis (Valent and Young, 1995).

Some waste sites will require action to remediate environmental damage and prevent further damage, either by recovery for reprocessing, disposal on land, or entombment in place. Remediation programs, especially those involving radioactive wastes, could become quite large and require extensive monitoring to which ROVs could contribute. Possible candidate sites include shallow, high-level radioactive waste sites in the Russian Arctic; shallow, low-level sites off the coast of California (Booth et al., 1989); and the Russian Arctic rivers. Other candidates include sunken nuclear submarines and nuclear weapons lost in deep waters. The Russian DSVs *MIR I* and *MIR II* have been used to inspect the site of the sunken Russian nuclear submarine *Komsomolets* and to cover the open torpedo tubes to prevent current flow from spreading radiation that may escape from the submarine's nuclear weapons (Seymour et al., 1994; OTA, 1995).

Existing undersea vehicles and techniques could be used to survey prospective sites for emplacement of waste such as sewage sludge. ROVs, not unlike those at work in the offshore oil and gas industry, could support much of the emplacement process, although such tasks could require greater depth capability than these systems currently provide. AUVs might perform this work more efficiently, but most of the survey requirements are within the capabilities of existing DSV and ROV technology or the near-term extensions of current capabilities. In any case, improved in situ sensing would be helpful if not necessary. Some surveys, however, could require capabilities that existing technology and existing service companies cannot presently supply.

Monitoring Water Quality

Dinoflagellate and algae blooms, often associated with water pollution in the coastal zone, can have major impacts on fisheries and coastal recreation. Toxic dinoflagellate or diatom blooms (often called "red tides") can result in shellfish poisoning, with far-reaching effects throughout the food chain. The "brown tide," which seems to be caused by the introduction of new detergent additives (Dzurica et al., 1989; Sieburth, 1989), caused the collapse of the scallop fishery in the middle-Atlantic and New England states, with economic losses of as much as $2 million per year.

To date, studies of coastal pollution and monitoring have not made wide use of undersea vehicles, although such use is growing. Scientists have used ROVs to make visual assessments of the effect of oxygen depletion on benthic and water column populations, to measure oxygen, and to collect plankton samples (Rabalais et al., 1991, 1992). In the future, undersea vehicles could make graphic or video surveys of the bottom near outfalls to assess population changes and diseased organisms. They could also monitor waste plumes or the abundance of toxic bloom organisms. An AUV system is also being developed that could monitor a local habitat on a regular basis from a moored garage (von Alt and Grassle, 1992; von Alt et al., 1994).

Much of the potential for using undersea vehicles in water-quality research and monitoring will depend on the development of sensors capable of high-quality measurements of environmentally important chemical species. Sensors designed for oceanographic work may prove useful in detection of dissolved trace metals and nutrients, but many important chemical species, such as chromium, cannot yet be detected using sensors suitable for undersea vehicles. Recent advances in biochemical sensors, including the development of DNA probes for organisms that produce the highly toxic domoic acid, have direct application for water-quality monitoring (Scholin et al., 1994). Biochemical sensors that can be deployed aboard undersea vehicles are rapidly being developed.

MARINE INDUSTRIAL ACTIVITIES

The sea provides a variety of nonliving resources, such as energy sources, metallic ores, and other minerals, and provides a venue for economic activities, such as developing and maintaining telecommunication linkage and undertaking salvage operations. In all of these areas, undersea vehicles offer unique tools for conducting inventories and extracting resources and for various inspection and manipulation tasks.

Oil and Gas

Offshore oil and gas production provides a major part of the U.S. energy supply—nearly one-fourth of the nation's natural gas and one-sixth of the domestic oil supply. While the market price for gas has rebounded recently to mid-1980 levels, world crude oil prices remain depressed; thus increasing pressures to reduce production costs. The restrictions on U.S. east and west coast offshore exploration and development imposed by Congress on environmental grounds leave the Gulf of Mexico as the focus for developing new fields and enhancing production from older ones. For the most part, the new fields of the Gulf are in water depths beyond the reach of divers, and their development requires using alternative technologies such as floating production facilities and undersea vehicles (Cafarelli, 1994). The rapidly changing environment for offshore oil and gas operations in the Gulf of Mexico—the increasing depth of new fields and greater distances between surface facilities—demands new tools and techniques for inspection and other tasks. Undersea vehicles have assumed a support role in many of these new approaches.

The offshore oil and gas industry is the leading user of undersea vehicles, worldwide, both in numbers and in dollar value. The industry used DSVs briefly during the early rapid growth of offshore exploration in the 1960s and 1970s, but ROVs—with their lower risk to humans, reduced system costs, and greater endurance—eventually displaced DSVs for most tasks. The proliferation of ROVs continued as exploration and production operations moved into waters deeper than 300 meters—beyond human diver capabilities—in the mid-1970s. That evolution shaped the present state of ROV development throughout the world to the extent that ROVs routinely perform many functions in offshore oil operations. While research and development for various sensors, tools, and subsystems originally came from other industries, the oil industry and its suppliers brought the technologies together and built practical working hardware. More than 90 percent of the nearly 300 nonmilitary, work-class ROV systems in use worldwide today are supporting offshore hydrocarbon exploration and production. The committee estimates that 80 percent of these systems are operating in foreign waters. The focal project on subsea oilfield service inspection and intervention (see Box 3-3) illustrates how presently available technologies can be focused to respond to the new requirements of the offshore industry in U.S. waters. The selection of an AUV over an ROV for such an application will depend on the relative costs and benefits of the two vehicle types. Cost evaluations should take into account support systems and flexibility in performing multiple tasks. An AUV's independence from surface support may have advantages in regions where weather and sea conditions can interfere with routine operations.

Marine Minerals

Mineral deposits on and beneath the EEZ seafloor of the United States are potentially an important national asset. A variety of metal ores are widely distributed there, including phosphorites at depths up to 4,500 meters, metalliferous sulfides (polymetallic sulfides) at depths greater than 700 meters, and metalliferous oxides (manganese nodules and crusts) at depths greater than 1,500 meters. For economic reasons, however, marine mining today is mainly a matter of recovering easily accessible minerals, such as sand and gravel or placer deposits in nearshore waters less than 100 meters deep, using readily available and cost-effective dredging techniques.

The recovery of deep water minerals will require new techniques, but there is little incentive for government or industry to invest in developing the technology. Depressed mineral prices and questions about the impacts of uncertain political and environmental policies (e.g., leasing, environmental effect, and restoration) will continue to slow development and exploitation for at least the next decade. Deep water mining is likely to begin sometime in the next 10 to 30 years, depending on market conditions (NRC, 1989).

In the near term, the industry needs scientific information about the extent and value of potential mineral resources in the EEZ. At present exploration is mainly a matter of mapping seafloor depths using shipboard and towed sensors. While this topographical mapping is useful, it provides little information on the extent or thickness of ore-bearing zones or on the concentrations of minerals within the area. Assessing seabed minerals quantitatively, with realistic economic valuations, will require new exploration techniques and sensors that can characterize and assay the constituents of ore deposits both in wide-area surveys (over hundreds of square kilometers) and in precisely localized measurements and samples. The committee's focal project on search and survey (see Box 3-4) illustrates a near-term application for a vehicle that could characterize seabed areas for minerals distribution.

To enable the recovery of deep water minerals, mobile seabed miners will likely be more practical in the longer term than the dredges used today in shallow water. Such mobile miners are essentially large ROVs, powered and controlled from surface vessels. They will perform tasks such as breaking up consolidated seabeds, scooping or dredging surface materials to precise depths below the seafloor, crushing and mixing the materials into slurries, and pumping or lifting the product to the surface. Mobile miners will require significant amounts of surface-supplied power (hundreds to thousands of kilowatts), seabed mobility and stability for a wide variety of slopes and bottom conditions, and accurate navigation to measure the quality and quantity of ore and minimize recovery of noncommercial product. When this type of activity does become practical, environmental monitoring of the area around the mining sites may be required as well, probably involving other types of undersea vehicles. To date, many mobile miner concepts have been proposed, and prototype systems have been tested, but no full-scale commercial equipment has emerged.

Ocean Energy

Electrical and chemical energy from the ocean could be practical alternatives to fossil fuels in some cases. Geothermal, ocean thermal, kinetic (wave or current), and biomass energy from the ocean can all be converted to useful forms. Several tidal and wave power plants have been built or are planned. Ocean thermal energy conversion (OTEC) has been demonstrated several times over the past century. OTEC may be the most practical ocean energy technology for large-scale production and may have the most relevance to undersea vehicles. OTEC converts the temperature differential between solar-warmed surface water and cold deep water (600 to 900 meters) into electrical energy using a circulating working fluid that is vaporized by warm surface water and then expands to operate a gas turbine before being recondensed by cold deep water. Low fossil fuel prices have combined with the high capital costs of OTEC plants to slow development since the 1970s. OTEC plant ships[4] of modest size (5 to 20 megawatts) are now under serious consideration for generating electricity in tropical and semitropical locations on islands and around the Pacific Rim, where both the geography and economics are favorable.

The Natural Energy Laboratory of Hawaii is building an experimental OTEC plant at Keahole Point, Hawaii. This shore-based facility is scheduled to be completed in 1996. Installation of a commercial plant in the next decade may lead the way to expanded use of the immense thermal energy stored in the world's oceans. Particularly appealing are the prospects for using the electrical output of an OTEC plant for production of combustible transportation fuels (Avery and Wu, 1994). The electricity is used on board the plant ship to hydrolyze seawater, which is then reacted with atmospheric nitrogen to produce ammonia or with carbon sources (such as coal) to produce methanol. Both ammonia and methanol can be burned in modified internal combustion engines. Secondary output includes gaseous hydrogen, oxygen, desalinated water, and other byproducts.

Undersea vehicles will play a role in the survey, construction, and operation of OTEC facilities. These functions will include exploration and survey of the seabed, where moorings and cold water pipe routes are planned and where data need to be acquired for environmental assessment and hardware design. Undersea vehicles may also perform inspection and maintenance of the subsea elements of commercial plants once they are installed. For example, ROVs would be used in much the same way as offshore oil platforms, performing inspection and maintenance tasks. An OTEC plant ship has a very large cold-water pipe reaching to approximately 900-meter depth. Inspecting and servicing this pipe is an important potential application for undersea vehicles; cleaning algae and barnacles from subsurface heat exchangers is another.

Mission requirements for undersea vehicles in the ocean energy industry would be similar to those in the offshore oil and gas industry, with the addition of specific sensors and intervention requirements. These requirements will involve a great deal of detailed observation and intervention, with manipulation capabilities of sufficient strength for in situ construction and maintenance tasks.

Salvage and Recovery

Salvors' traditional use of divers had restricted work to depths of less than 100 meters until the past 30 years. Advances in diving technology begun in the 1960s, such as saturation diving and modern diving suits, have since extended salvage capabilities to greater depths, depending on the specific task. While undersea vehicles can be used to salvage materials for their commercial value, recovery can also make important contributions that affect public safety and national security. Recovery of a jet engine, for example, rarely yields hardware financially valuable except as scrap; however, such an engine may hold clues to the causes of a plane crash. For example, the U.S. Navy has used DSVs and ROVs to recover F-14 engines, which were then examined to determine the cause of engine failure to prevent further accidents. Undersea vehicles are used routinely to recover entire aircraft, flight recorders, hazardous materials, and other materials requiring study and analysis.

The Navy first developed and used ROVs for recovering torpedoes at test sites. In 1966, after an extensive search by various types of undersea vehicles and finally after it was located by the DSV *Alvin*, a U.S. Navy ROV, a cable controlled underwater recovery vehicle (CURV), recovered a lost hydrogen bomb from a depth of nearly 1,000 meters. In 1986, the Navy's nuclear-powered research submarine, *NR-1*, located and classified debris from the space shuttle *Challenger*.[5] Critical components of the solid rocket booster were then verified by the DSV *Johnson Sea-Link I* and recovered by the ROV *Gemini* (NAVSEA, 1988). In 1990, the DSV *Sea Cliff* recovered the cargo door of a United Airlines 747 from a depth of 4,317 meters in the Pacific, following search and location by the *Orion* search system. This recovery and others, such as recovery of components of an Air India 747 lost in the North Atlantic off Ireland and of a South African Airlines 747 that crashed into the Indian Ocean off Mauritius, have caused the public to expect the federal government to use national assets, as necessary, to provide clues to aircraft and maritime accidents and unexplained losses.

Treasure hunters are using undersea vehicles more

---

[4]Plant ships are floating structures containing all elements of an OTEC plant. A plan tships may be a grazer or may be moored in a fixed position.

[5]For 7 months, beginning February 8, 1986, a search was conducted under the supervision of the U.S. Navy leading to the salvage of *Challenger*. The search involved a systematic inspection of approximately 1,666 km^2 (643 mi^2), in depths ranging from 3 meters to over 370 meters. This operation was supported by four ROVs and two DSVs. This was the largest search and salvage mission ever undertaken in terms of geographic area, weight, and number of pieces salvaged (NAVSEA, 1988).

> **BOX 3-3**
> **Focal Project 3: Subsea Oil Field Inspection and Intervention**
>
> **Opportunity**. In an effort to reduce the cost of hydrocarbon production in the deeper waters of the outer continental shelf, the U.S. oil industry is turning from fixed or floating platforms to "subsea completion systems." With this technology, a "satellite" unit operates on the seafloor with a platform of its own. These satellite installations include seafloor templates and wellheads and are within oversight proximity to a central production platform, which can be up to 46 km (25 nautical miles) away. Each production platform can handle a number of satellites. Satellite units direct the oil or gas through a pipeline along the seabed to the central platform.
>
> **Objective**. Satellite installations and their pipeline connections require periodic inspection and intervention for continued, reliable operation. Presently this involves leasing a support vessel outfitted with an ROV system, with the possible addition of a towed sensor to inspect flowlines and control cables. An AUV could perform satellite and pipeline inspection and intervention tasks by operating from the central platform and making excursions to the wellhead sites. The AUV would eliminate the need for leasing an expensive surface vessel (at approximately $20,000 per day), and it would be able to survey pipelines and control cables linking the central platform with the satellite subsea wellheads while in transit to or from the satellite. Response to unplanned and emergency situations would be more rapid than using surface vessels traveling from a shore service site.
>
> **Vehicle System**. The accompanying figure illustrates the operations concept of an oil field service AUV, which would be "garaged" at the central platform and would be capable of operating autonomously or under supervisory control of an operator on the platform. At the satellite, the vehicle would connect to a fiber-optic link permanently installed in the satellite for operation and feedback in the supervisory mode. The vehicle could perform detailed observation and intervention of the satellite with an operator in the control loop in real time, if the system had task-level control.[1] The vehicle would then transit back to the platform autonomously; it could also return on its own at any time during the mission in case of a loss of communications. If multiple sorties were required, the vehicle could be recovered and replenished by recharging the batteries.
>
> **Performance Requirements**:
> - Range: Transit distance, 93 km (50 nautical miles) round-trip
> - Speed: 6 kt (maximum)
> - Mission Duration: 36 hours (maximum)
> - Sensing Capability: Video, acoustic, magnetic, line scanning
> - Work Package: Manipulators with special end-effectors for rotating valves, cutting and clearing debris, removing and replacing plug-in modules
> - Control: Task-base
>
> ---
> [1] Under the supervisory control mode of operation, a telemetry link in the form of an acoustic modem or a lightweight fiber-optic tether would provide a real-time connection from the AUV to the platform.

frequently in their salvage efforts. In the early 1990s, the Columbus America Discovery Group used an ROV to recover gold worth millions of dollars from the ship SS *Central America* at a depth of more than 2,000 meters. The Monterey Bay Aquarium Research Institute used an ROV to photograph and survey the sunken Navy dirigible *Macon*. Similar operations were performed on the *Titanic*, *Bismarck*, and *Lusitania*. Such efforts are neither pure archaeology nor pure salvage, but combinations thereof.

Requirements for salvage include both large-area surveys and detailed observation and manipulation. DSVs and ROVs have repeatedly performed these tasks successfully. AUVs, such as *AUSS*, the U.S. Navy Advanced Unmanned Search System vehicle, have also performed automated search and survey operations. This capability, widely applied, could significantly improve the efficiency and success rate of salvage and recovery operations, especially if employed in conjunction with DSVs or ROVs.

For many years, marine archaeologists have used divers to work in relatively shallow waters. During the past four decades, pioneers, including Edwin Link, Willard Bascom, and George Bass, used undersea vehicles for archaeology at much greater water depths, adapting technology developed mainly for oceanographic research, offshore oil operations, and military activities. In 1966, the DSV *Aluminaut* discovered two cannon-carrying wooden ships on the seafloor off Palomares, Spain; in 1974, a *Johnson Sea-Link* DSV surveyed the wreck of the

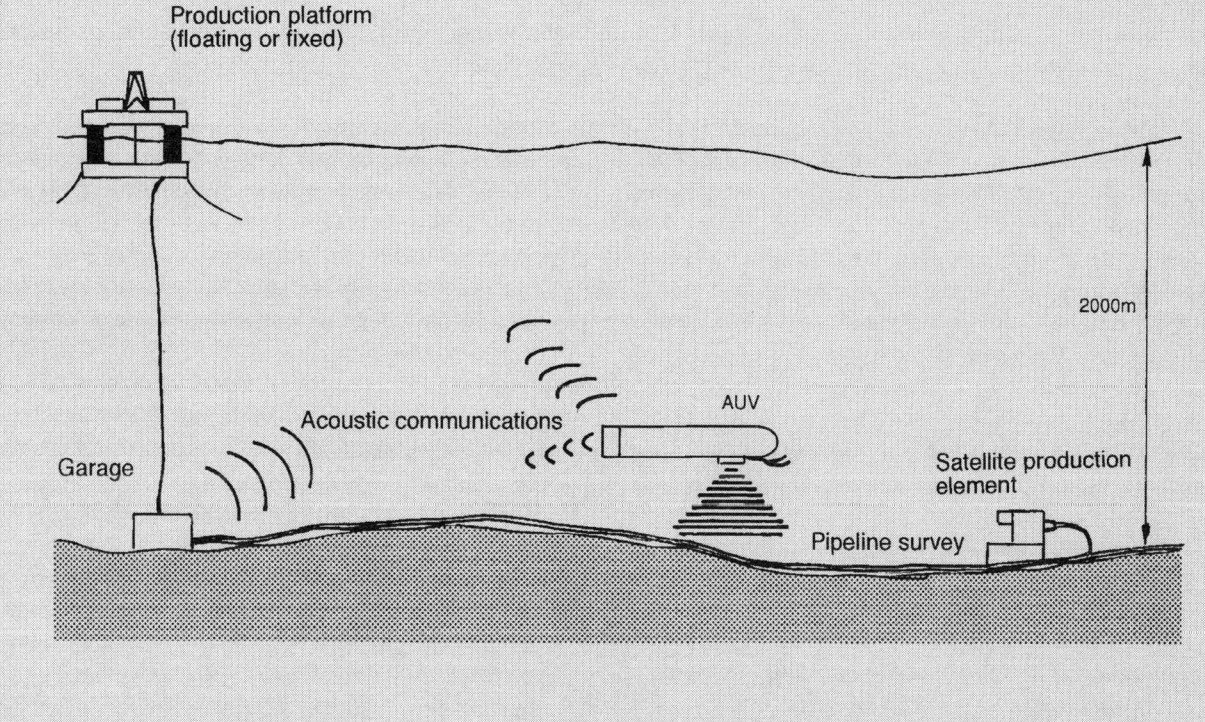

Analysis of the State of Technology and Practice. The basic technology supporting the AUV vehicle design and operation is in place. However, development will be required in the following areas:

- adaptation of advanced batteries such as those having secondary (rechargeable) lithium chemistries
- integration of the sensors, navigation, and control subsystems
- adaptation of magnetic sensors to AUV operation

To accomplish intervention tasks at the subsea satellite, improved electronically operated manipulators will be required, and special features will have to be built into the satellites to permit the most efficient use of service AUVs.

Subsea Oil Field Inspection and Intervention Vehicle Concept

USS *Monitor*, producing a photo mosaic; and 1990, archaeologists used an ROV to examine and photograph *Hamilton* and *Scourge*, American warships sunk during the War of 1812, on the bottom of Lake Ontario.

An archaeological survey and excavation must meet exacting standards, using procedures that maximize the amount of information available from a site. Much of this work has primarily been observational because there is neither the technology nor the procedures to perform proper archaeological excavations below diver depths. Extensions of undersea salvage capabilities combined with archaeological survey standards could move archaeology surveys into deeper depths. Many of the required capabilities may also be adapted from those developed for other applications, such as oceanographic surveys and sampling, and from scientific, military, and commercial search and survey methods.

The focal project on search and survey (see Box 3-4) describes the potential for using a conceptual, improved AUV optimized for performing the search and survey tasks that must precede recovery or salvage operations.

Communications

One of the world's fastest growing industries is submarine telecommunications. Lightweight fiber-optic cable, with its great bandwidth, clarity of transmission, and nonrepeater systems that can span long distances (>300 kilometers) is

BOX 3-4
Focal Project 4: Search and Survey

Opportunity. Full-ocean search and surveys to 6,000 meters for sunken objects or valuable materials on the seafloor are currently undertaken with towed systems that require winches and cables. The cost of towed search equipment, support ships, boats, and personnel is significant—exceeding $40,000 per day for operations over 5,000 meters depth. Moreover, search techniques using towed bodies require over half of the operational time to perform turns in the search pattern.

In situ observation during search operations and the ability to provide quick, detailed search and identification response to probable targets are needed to reduce cost and search time. A method of providing limited object retrieval capability will further reduce overall operational costs and reduces the probability of loss of target during the search mode. In addition, portability and flexibility of the search and survey system will be enhanced if the system is not dependent on a dedicated surface support ship and if it can be used on a ship of opportunity.

Objective. To survey and locate seabed minerals such as nodules, sunken vessels, high-value debris, archeological objects, and aircraft wreckage, and to recover small high-value objects.

Vehicle System. An AUV, like the one shown below, can be launched from a platform crane or stern ramp of a ship of opportunity. After launch, the AUV dives to the seafloor and executes a preprogrammed, long-range side-scan search or survey while traveling approximately 30 meters above the seafloor. The vehicle may carry opportunistic sensors (nonintrusive mineral detectors and subbottom profilers to detect ore bodies or buried materials of value). The vehicle then performs a closer inspection and identifies any targets, using higher-resolution sonar and optical video imagers. The vehicle recovers an identified object with its own manipulator or attaches a lift line to the object for recovery by a surface vessel. Both tasks require heavy and precise manipulation capabilities. The AUV then returns to the surface on command at the end of the survey or recovery.

The mission involves searching and discovery. Human designation of tasks in response to opportunities is called for. Humans monitoring the preprogrammed search in near real time can assume low-bandwidth, task-level control when desired, owing to recent advances in acoustic communications and high-level control architectures (as in focal projects 1 and 2).

Performance Requirements:
- Coverage for search: 300 km^2 per day.
- Depth: 6,000 meters (maximum)
- Speed: 0 to 5 kt
- Length: 12 meter (2 meter modules), estimated
- Displacement: 11,300 kg (about 25,000 lb)
- Position: Line following and way-point processing
- Energy: 200 kWh
- Sensors:
    — obstacle avoidance sonar
    — swath, side-scan sonar
    — laser-line scanner
- Work Packages: Modules for specific applications
    — sensors for minerals survey
    — manipulator module
    — coring module
- Communication: Acoustic telemetry, real time over a single mode fiberoptic link (10,000 meters fiber-optic spool)

Analysis of the State of Technology and Practice. DSVs and ROVs have routinely performed search and recovery tasks successfully. Over the past decade, several types and sizes of AUVs, including the Navy's Advanced Unmanned Search System (*AUSS*) vehicle, have been fielded carrying a variety of sensor suites. These have mostly operated in depths less than 1,000 meters. Continuing developments in control, data processing, data storage, acoustics, and optics have enhanced the feasibility of using large AUVs for these missions, but substantial advances will be needed in the following areas:

- advanced batteries with energy density in the range of 200 to 400 kWhr/kg
- low power swath sonar
- low power obstacle avoidance and path planning subsystem
- low power laser line scanner
- terra-byte data storage
- power and data storage development to extend mission endurance

Search and Survey Concept

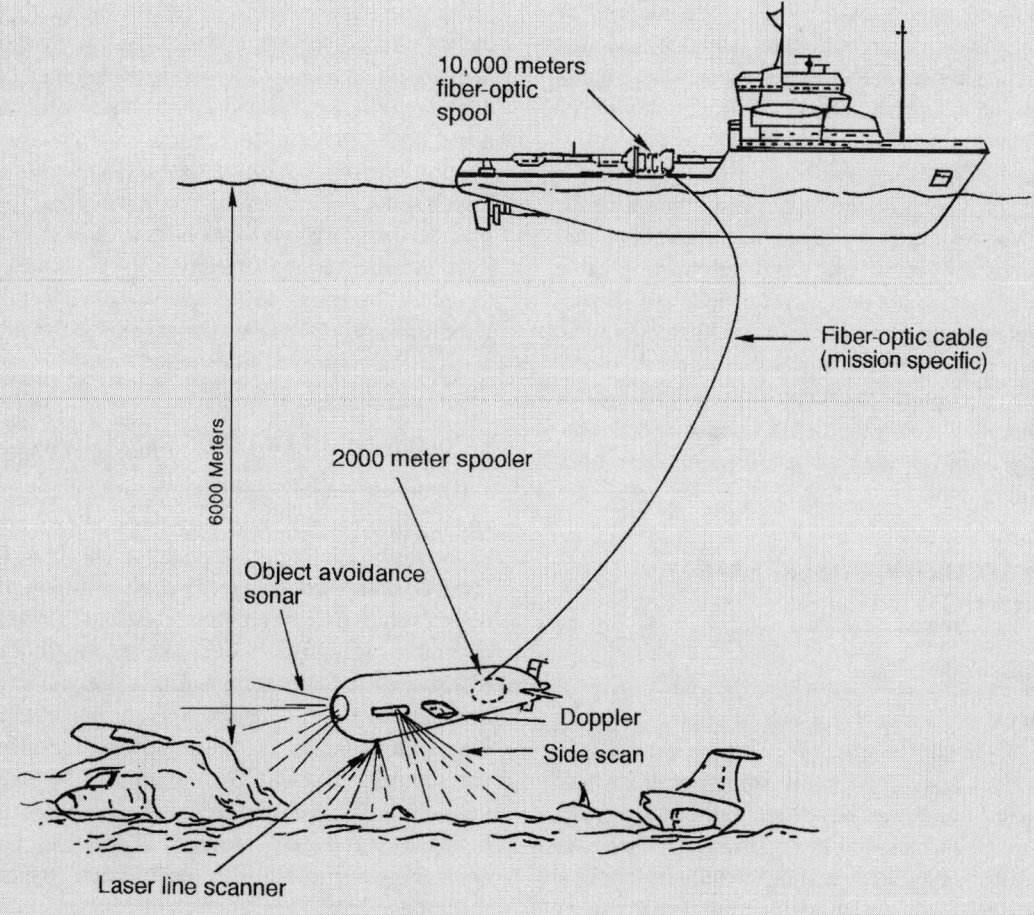

becoming the mainstay of global communications. Systems currently being installed provide 320,000-channel capacity, 40 times greater than could be achieved 15 years earlier. Construction of undersea cables is a nearly $1 billion enterprise, growing 7 to 8 percent per year. Between $8 billion and $10 billion of undersea fiber-optic systems will be built by late 1997, according to American Telephone and Telegraph. These systems will crisscross the globe, with long runs across the Atlantic and Pacific and many shorter runs linking islands and coastal cities (Keller, 1992; Bannon, 1993).

An important part of this expansion, especially in terms of construction and maintenance, is the interconnection of coastal cities. Problems in obtaining land rights-of-way have led some telecommunications systems to connect cities by sea where possible, using short-haul, repeaterless systems along the coastlines. These systems require cable to be buried to depths greater than about 350 meters in areas of commercial shipping or fishing. Assurance of the integrity of these rapidly expanding undersea interconnections requires the ability to repair cables in the deeper water, where faults have occurred due to human activity, sediment slumps, or earthquakes. The current method of maintaining connectivity in these cables, when interrupted by fishing or seismic activity, is the recovery, assisted repair, and reburial of the cables by ROVs. An example is American Telephone and Telegraph's advanced submersible craft for assisting cable repair and burial, which are designed for undersea inspection, maintenance, and repair operations and can operate in rough seas at depths up to 2,500 meters (Joggerst, 1995). These vehicles will continue to occupy a specialized niche in advanced technology within the field of undersea vehicles, largely developed and operated in house by communications companies (Smith, 1992).

NATIONAL SECURITY, PUBLIC SAFETY, AND REGULATION

Military Applications

The military was the main force behind the development of undersea vehicle technology in the United States. Techniques and devices developed for military applications have led the field in technology and have been transferred to other fields, such as offshore oil and gas operations. They have also been used for scientific research for military applications. Oceanography and ocean engineering research have been central to the U.S. Navy's support of DSVs. Navy-sponsored DSVs have tested materials and sensors and explored the ocean for military and scientific users. They have also been used for submarine rescue; deep-ocean search (to 6,000 meters); and location, inspection, and retrieval of items on the seabed. Navy assets include three, three-person DSVs: *Sea Cliff* (6,000 meters), *Turtle* (3,000 meters), and *Alvin* (4,500 meters, operated by the Woods Hole Oceanographic Institution); two research submarines, *NR-1* (nuclear powered) and USS *Dolphin* (battery/diesel powered); and two deep-submergence rescue vehicles, *Mystic* and *Avalon*, whose hatch skirts are capable of mating to U.S. and foreign submarines.

The Navy began development of ROVs in the 1960s to perform many of the tasks of DSVs at lower cost and with greater safety. However, DSVs continued to be the dominant types of vehicles used in research, search, and high-value object retrieval well into the 1970s. Heavy-duty work ROVs, such as *CURV III* and the advanced tethered vehicle (6,000 meters), can perform deep-ocean work such as the retrieval of missiles and torpedoes and the salvage of military aircraft. Other military ROVs can inspect, implant, and recover objects, salvage, and locate and neutralize mines.

During the past decade the military has developed or sponsored the development of both AUVs and semiautonomous vehicles. AUVs could multiply the effectiveness of today's smaller military forces. Potential applications include search, detection, and identification; weapon targeting and placement; countermeasures employment (mines, torpedoes, sonar); mapping and survey; communications; surveillance; and object implantation and recovery. Military-supported AUVs include the *AUSS*, which can perform search and survey to depths of 6,000 meters. The *Unmanned Free-Swimming System* was built to demonstrate the viability of low drag, long-distance AUVs. In addition, as noted in Chapter 1, two large ARPA sponsored AUVs built by Draper Laboratories have served as testbeds for mine countermeasures and other military missions.

Enforcement of Laws and Regulations

Undersea vehicles aid in enforcing a wide variety of laws and regulations—from helping detect illegal pollution to interdicting illegal drug shipments. The U.S. Coast Guard (USCG) is the federal agency that most frequently uses undersea vehicles for nonmilitary missions. Three major USCG programs presently use or could benefit from undersea vehicles: search and survey, marine environmental protection, and oil and toxic chemical spill response (USCG, 1994). Undersea vehicles could also perform surveillance for law enforcement. The USCG is responsible for enforcing international agreements and federal laws over, on, and under the high seas and waters subject to U.S. jurisdiction. These responsibilities include drug interdiction, fisheries enforcement, and illegal immigration interdiction.

Other uses for undersea vehicles are monitoring for pollutants, sampling beneath ships entering and leaving ports to ensure compliance with applicable laws, and conducting surveillance and under-hull inspections of vessels suspected of carrying contraband. The USCG's marine environmental protection program includes two major missions: marine environmental response and port safety, both of which respond to pollution threats in coastal waters and some inland waters. In support of the 1972 Federal Water Pollution Control Act, the USCG prepared a national oil and hazardous substance spill response system that requires contingency planning and interagency and international relationships. In many areas ROVs are part of the regional response systems assets. The USCG uses undersea vehicles to respond to oil and toxic chemical spills in coastal waters. For example, ROVs are used to evaluate leaking vessels damaged by groundings or collisions.

Real-time assessment of damage below the waterline can speed responses to accidents and limit damage to the environment. A vehicle assessing below-waterline damage to ships would perform a detailed observation and intervention task. Such a vehicle would need sonars and optical/video sensors to record damage and would have to be compact, highly maneuverable, and readily transportable to accident sites. An ROV might attach to the vessel and crawl along the hull to assess damage and possibly to aid in plugging or patching holes. The ROV would need dexterous manipulators capable of handling a variety of tools performing underwater cutting and welding. These ROVs could also aid in responding to pollution hazards by performing medium-area (tens of kilometers) surveys with chemical sensors to assess the extent of a spill. Vehicles performing law enforcement tasks, such as searching for contraband or providing port security, would operate in shallow water with acoustic sensors, performing local area or dispersed surveys.

FINDINGS

Finding. Humanity's need to explore and work in the ocean is growing day by day. Coastal populations are increasing worldwide, and with them, both coastal pollution and dependence on ocean resources such as fish. Nonliving resources (including oil, gas, and other minerals) require exploration and extraction on and beneath the seafloor in deeper and deeper water. Profound discoveries in the Earth sciences—about the climatic future and the planet's tectonic structure—continue to be made through undersea activities. Defense and law enforcement agencies have their own growing requirements for undersea vehicles. All of these national needs can be met more effectively through developing the capabilities of undersea vehicles.

Finding. Scientific exploration of the seafloor with undersea vehicles has given scientists firsthand evidence of seafloor spreading and other processes of plate tectonics that have immense significance for science and for the future of humanity. Findings made possible by undersea vehicles offer fundamental geological and biological knowledge of great value. Undersea vehicles also provide strategic and economic benefits, helping locate resources and predicting volcanic activity and earthquakes.

Finding. Undersea vehicles will allow more accurate measurements of the flows of chemicals and heat in the ocean. For example, they can be used to acquire data to determine the extent to which the greenhouse gas, carbon dioxide, is removed from the atmosphere and sequestered in the deep oceans.

Finding. The biological productivity of the oceans is only dimly understood today owing to the scarcity of data on ocean food chains. Today's scientists rely largely on towing nets behind surface vehicles to sample many kinds of ocean life. Undersea vehicles can give scientists direct access to the plants and animals, including many that are thought to be too delicate or elusive to be captured by nets. The hitherto unsuspected existence of specialized communities at hydrothermal vents and cold seeps was confirmed by scientists using undersea vehicles. Further exploration is certain to reveal other profound discoveries, including, perhaps, new biological chemicals with medical or industrial uses.

Finding. Better management of fisheries is vital to continue feeding the Earth's growing population. Many of the world's commercial fisheries are overexploited. Undersea vehicles could directly collect information on rates of birth, growth, and death; predator-prey relationships; spawning and nursery ground requirements; and the impact of intensive fishing.

Finding. The ocean contains a large, but poorly quantified and characterized, amount of waste that is both deliberately disposed of and inadvertently allowed to reach the water through accidents and runoff from land. Undersea vehicles with new kinds of chemical sensors could help detect pollutants and trace them to their sources. Conventional ROVs, with video cameras and sampling tools, are already used to monitor disposal sites. Others may be useful in assessing the wisdom of waste disposal in deep sea trenches or on the abyssal plains.

Finding. Finding, developing, and supporting routine production of the ocean's nonliving resources—particularly offshore oil and gas—has involved a number missions for ROVs (such as manipulation of values and controls, cutting, replacement of components, inspection, and monitoring), many of which require capabilities to perform increasingly complex functions. Mounted video cameras and magnetic sensors are used routinely in the offshore oil and gas industry. ROVs are now being used to support production in depths to 1,000 meters and have been used in exploratory drilling at more than 2,000 meters. Deeper water, more complex applications, and more wells will require the continued evolution of ROVs, and perhaps other vehicles, with more fully automated devices, better navigation and location systems, and greater capabilities for operation and maintenance tasks. The metals and other minerals found on the seabed will require new surveying and prospecting techniques and powerful undersea dredges and loaders should extraction become fensible and economical. Another extremely important and rapidly growing market for ROVs is in laying telecommunication cables.

Finding. A variety of industrial needs can be met by undersea vehicles. Salvage and retrieval of valuable items from the seafloor is one of the most well-established applications of undersea vehicles. This application originated in the U.S. Navy programs of the 1960s, and undersea vehicles proved themselves in missions such as the after-crash recovery of aircraft engines and other components.

Finding. Military applications (including rescue, search and retrieval, and mapping of the bottom) are served by a small fleet of DSVs, ROVs, and AUVs. Since mid-1994, several of the DSVs and an ROV have been made available for scientific research when they are not committed to defense missions or would otherwise be idle. The U.S. Navy and ARPA are the sources of most research and development support for AUVs.

Finding. Law enforcement and regulatory agencies use undersea vehicles in a variety of missions, including helping detect pollution, interdicting illegal drug shipments, and monitoring marine sanctuaries and other protected areas. Improved chemical and optical sensors will make all of these tasks easier and more effective.

REFERENCES

Adams, M.W.W., F.B. Perler, and R.M. Kelly. 1995a. Extremozymes: Expanding the limits of biocatalysis. Biotechnology 13 (7):662–668.

Adams, P.B., J.L. Butler, C.H. Baxter, T.E. Laidig, K.A. Dahlin, and W.W. Wakefield. 1995b. Population estimates of Pacific coast groundfishes from video transects and swept area trawls. Fishery Bulletin 93: 446–455.

Avery, W.H., and C. Wu. 1994. Renewable Energy from the Oceans: A Guide to OTEC. New York: Oxford University Press.

Bailey, T.G., J.J. Torres, M.J. Youngbluth, and G.P. Owen. 1994. Effect of decompression on mesopelagic galetinous zooplankton: A comparision of in situ and shipboard measurements of metabolism. Marine Ecology–Progress Series 113(1–2):13–27.

Bannon, R.T. 1993. Cable protection and preventive maintenance in Micronesia and Polynesia. Pp. 105–108 in the Proceedings of the 11th Annual Conference, Underwater Intervention '93 held January 18–21, 1993 in New Orleans, Louisiana. Washington, D.C.: Marine Technology Society.

Booth, J.S., W.J. Winters, L. Poppe, J. Neiheisel, and R. Dyer. 1989. Geotechnical, geological, and selected radionuclide retention: Characteristics of the radioactive waste disposal site near the Farallon Islands. Marine Geotechnology 8:11–132.

Bothner, M.H., H. Takada, J.T. Knight, R.T. Hill, B. Butman, J.W. Farrington, R.R. Colwell, and J.F. Grassle. 1994. Sewage contamination in sediments beneath a deep-ocean dump site off New York. Marine Environmental Research 38:43–59.

Cafarelli, W.J. 1994. Possible uses of autonomous underwater vehicles in offshore seismic exploration. P. 11 in MIT Marine Industry Collegium Symposium on Commercialization of Underwater Vehicles held January 25–26, 1994 in Cambridge, Massachusetts. Report No. MITSG 93-32. Cambridge, Massachusetts: MIT Sea Grant Program.

Calmet, D.P., and J.M. Brewers. 1991. Radioactive waste and ocean dumping: The role of the IAEA. Marine Policy 15(6):413.

Coale, K.H., C.S. Chin, G.J. Massoth, K.S. Johnson, and E.T. Baker. 1991. In situ chemical mapping of dissolved iron and manganese in hydrothermal plumes. Nature 352:325–328.

Corliss, J.B., J. Dymond, L.I. Gordon, J.M. Edmond, R.P.V. Herzen, R.D. Ballard, K. Green, D. Williams, A. Bainbridge, K. Crane, and T.H. Vanandel. 1979. Submarine thermal springs on the Galapagos Rift. Science 203:1073–1083.

Curtin, T.B., J.G. Bellingham, J. Catipovic, and D. Webb. 1993. Autonomous oceanographic sampling networks. Oceanographics 6(3):86–93.

Davis, C.S., Gallager, S.M., and Solow, A.R. 1992. Microaggregations of oceanic plankton observed by towed video microscopy. Science 257:230–232.

Dzurica, S., C. Lee, E.M. Cosper, and E.J. Carpenter. 1989. Role of environmental variables, specifically organic compounds and micronutrients, in the growth of the Chrysophyte Aureococcus anophagefferens. Pp. 229–253 in Novel Phytoplankton Blooms: Causes and Impacts of Recurrent Brown Tides and Other Unusual Blooms. E.M. Cosper, V.M. Bricelj, and E.J. Carpenter, eds. New York: Springer-Verlag.

Edmond, J.M., and K.L. Von Damm. 1979. Chemistry of hot springs on the East Pacific Rise and their effluent dispersal. Nature 297:187–191.

Flam, F. 1994. The chemistry of life at the margins. Science 265:471–472.

Fox, G.C. 1995. Special collection on the June 1993 volcanic eruption on the coaxial segment, Juan de Fuca Ridge. Geophysical Research Letters 22(2):129–130.

Fox, P.J., and D.G. Gallo. 1984. A tectonic model for ridge-transform-ridge plate boundaries: Implications for the structure of oceanic lithosphere. Tectonophysics 104:205–242.

Guinasso, N.L., I.R. MacDonald, J.M. Brooks, R. Sassen, and K.M. Scott. 1994. Seafloor gas-hydrates: A video documenting oceanographic influences on their formation and dissociation. Transactions in the American Geophysical Union (EOS) 75:175.

Hamner, W.M., and B. H. Robison. 1992. In situ observations of giant appendicularians in Monterey Bay. Deep Sea Research 39(7/8): 1299–1313.

Hawkes, G.S., and P.J. Ballou. 1990. The ocean everest concept: A versatile manned submersible for full ocean depth. Marine Technology Society Journal 24(2):79–86.

Hill, R.T., I.T. Knight, M.S. Anikis, and R.R. Colwell. 1993. Benthic distribution of sewage sludge indicated by Clostridium perfringens at a deep-ocean dump site. Applied Environmental Microbiology 59:47–51.

Jannasch, H.W. 1995. Deep-sea hot vents as a source of biotechnologically relevant microorganisms. Journal of Marine Biotechnology 3:1–8.

Joggerst, P. 1995. AT&T's undersea fiber-optic cable systems. Sea Technology 36(7):29.

Keller, J.J. 1992. AT&T racing rivals to lay undersea lines. *The Wall Street Journal*, August 14, 1992. B1 and B8.

Krieger, K.J. 1993. Distribution and abundance of rockfish determined from a submersible and by bottom trawling. Fishery Bulletin 91:87–96.

Krieger, K.J., and M.F. Sigler. 1996. Catchability coefficient for rockfish estimated from trawl and submersible surveys. Fishery Bulletin 94: 282–288.

Matsumoto, G.I., and G.R. Harbison. 1993. In situ observations of foraging, feeding, and escape behaviors in three orders of oceanographic Ctenophores: Lobata, Cestida, and Beroida. Marine Biology 117:279–287.

McDowell, S.E. 1994. Seafloor monitoring requirements of a major dredged material disposal project offshore New Jersey. P. 16 in MIT Marine Industry Collegium Symposium on Commercialization of Underwater Vehicles held January 25–26, 1994 in Cambridge, Massachusetts. Cambridge, Massachusetts: MIT Sea Grant Program.

NOAA (National Oceanic and Atmospheric Administration). 1993. 1995–2005 Strategic Plan, Build Sustainable Fisheries. (Available from Deputy Undersecretary, NOAA, Room 5810 Herbert Hoover Building, 14th and Constitution Avenue, N.W., Washington D.C. 20230.)

NAVSEA. 1988. Space Shuttle *Challenger* Salvage Report. NAVSEA T9597–AA–RPT/00. (Available from Commanding Officer, Navy

Publication and Forms Center, 5801 Tabor Avenue, Philadelphia, Pennsylvania 19120.)

New England Biolabs Catalog. 1994. DNA/RNA Modifying enzymes: Vent DNA polymerase and Deep vent DNA polymerase for high fidelity and thermostability during thermal cycle DNA sequencing. Beverly, Massachusetts: New England Biolabs, Inc.

NRC (National Research Council). 1989. Our Seabed Frontiers: Challenges and Opportunities. Marine Board, NRC. Washington, D.C.: National Academy Press.

NRC. 1993. Applications of Analytical Chemistry to Oceanic Carbon Cycle Studies. Committee on Oceanic Carbon. Ocean Studies Board, National Research Council. Washington, D.C.: National Academy Press.

NRC. 1995. Understanding Marine Biodiversity: A Research Agenda for the Nation. Committee on Biological Diversity in Marine Systems. Ocean Studies Board, National Research Council. Washington, D.C.: National Academy Press.

OTA (Office of Technology Assessment). 1995. Nuclear Wastes in the Arctic: An Analysis of Arctic and Other Regional Impacts from Soviet Nuclear Contamination. OTA-ENV-623. Washington, D.C.: U.S. Government Printing Office.

Piccard, J., and R.S. Dietz. 1960. Seven Miles Down. New York: G.P. Putnams.

Pilskaln, C.H., M.W. Silver, D.L. Davis, K.M. Murphy, B. Gritton, S. Lowder, and S. Lewis. 1991. A quantitative study of marine aggregates in the mid-water column using specialized ROV instrumentation. Pp. 1175–1182 in Volume 2 Proceedings Oceans '91 held October 1–3, 1991 in Honolulu, Hawaii. Pisactaway, New Jersey: IEEE Service Center.

Rabalais, N.N., R.E. Turner, D.E. Harper, Jr., and Q. Dortch. 1991. The use of remote and electronic surveillance to evaluate the Impacts of Hypoxia on Fisheries Resources. Pp. 5-77–5-80 in 1991 Annual Report submitted to NOAA's Office of Undersea Research by NOAA's National Undersea Research Center at the University of North Carolina. Wilmington, North Carolina: University of North Carolina.

Rabalais, N.N., D.E. Harper, Jr., R.E. Turner, and N.H. Marcus. 1992. Impacts of hypoxia/anoxia on living resources. Pp. 5-125–5-128 in 1992 Annual Report submitted to NOAA's Office of Undersea Research by NOAA's National Undersea Research Center at the University of North Carolina. Wilmington, North Carolina: University of North Carolina.

Robison, B.H. 1993. New technologies for sanctuary research. Oceanus 36(3):75–80.

Robison, B.H. 1995. Light in the ocean's midwaters. Scientific American 273(1):60–64.

Scholin, C.A., M.C. Villac, K.R. Buck, J.M. Krupp, D.A. Powers, G.A. Fryxell, and F.P. Chavez. 1994. Ribosomal DNA sequences discriminate among toxic and non-toxic *Pseudonitchia* species. Natural Toxins 2(4):152–165.

Seymour, R.J., D.R. Blidberg, C.P. Brancart, L.L. Gentry, A.N. Kalvaitis, M.L. Lee, J.B. Mooney, and D. Walsh. 1994. World Technology Evaluation Center Program. Pp. 150–262 in World Technology Evaluation Center Panel Report on Research Submersibles and Undersea Technologies. NTIS Report No. PB94-184843. Baltimore, Maryland: Loyola College of Maryland.

Sieburth, J.M. 1989. Are Aureococcus and other nuisance algal blooms selectively enriched by the runoff of turf chemicals? Pp. 779–784 in Novel Phytoplankton Blooms: Causes and Impacts of Recurrent Brown Tides and Other Unusual Blooms. E.M. Cosper, V.M. Bricelj, and E.J. Carpenter, eds. (Volume 35, Coastal and Estuarine Studies). New York: Springer-Verlag.

Smith, K.S., Jr. 1995. Scripps Institution of Oceanography. Personal communication to Dr. Mary Scranton, August 17, 1995.

Smith, T.J. 1992. SCARAB price performance—A study in efficiency. Pp. 57–61 in Proceedings of the 10th Annual Conference Intervention/ROV '92 held June 10–12, 1992, San Diego, California. Washington, D.C.: Marine Technology Society.

Stone, R. 1993. Déjà vu guides the way to new antimicrobial steroid. Science 259:1125.

Tucholke, B. 1995. Personal communication to Donald W. Perkins, November 9, 1995.

USCG (United States Coast Guard). 1994. Budget in Brief, Fiscal Year 1995. Washington, D.C.: U.S. Coast Guard.

UNOLS (University-National Oceanographic Laboratory System). 1994. The global abyss: An Assessment Deep Submergence Science in the United States. (Available from UNOLS, P.O. Box 392, Sanderstown, Rhode Island 02874 or e-mail: UNOLS@gso.uri.edu.)

Valent, P.K., and D.K. Young. 1995. Technical and economic assessment of storage on industrial waste on abyssal plains. Presented to Marine Board, National Research Council, June 21, 1995 at Stennis Space Center, Mississippi, May 25, 1995.

Van Dover, C.L., J.L. Grassle, B. Fry, R.H. Garritt, and V.R. Starczak. 1992. Stable isotope evidence for entry of sewage-derived organic material into a deep-sea food web. Nature 360:153–156.

von Alt, C.J., and J.F. Grassle. 1992. LEO-15: An unmanned long-term environmental observatory. Pp. 849–854 of the Proceedings of Oceans '92 held October 26–29, 1992 in Newport, Rhode Island. Piscataway, New Jersey: IEEE Service Center.

von Alt, C., B. Allen, T. Austin, and R. Stokey, 1994. Remote environmental measuring units. Pp. 13–14 of the Proceedings of the IEEE Symposium on Autonomous Underwater Vehicles AUV '94 held July 1-19, 1994 at Cambridge, Massachusetts. Piscataway, New Jersey: IEEE.

Von Damm, K.L., J.M. Edmond, B. Grant, C.I. Measures, B. Walden, and R.F. Weiss. 1985. Chemistry of submarine hydrothermal solutions at 21 N. East Pacific Rise. Geochimica Cosmochimica Acta 49:2221–2237.

Widder, E.A., S.A. Bernstein, D.F. Bracher, J.F. Case, K.R. Reisenbichler, J.J. Torres, and B.H. Robison, 1989. Bioluminescence in the Monterey Submarine Canyon: Image analysis of video recordings from a midwater submersible. Marine Biology 100:541–551.

4

Priorities for Future Development

This chapter presents the committee's evaluation of the subsystem technologies reviewed in Chapter 2 and identifies the technology investments with the greatest potential for advancing undersea vehicle capabilities for meeting the national needs and opportunities discussed in Chapter 3.

IMPORTANT NEEDS AND OPPORTUNITIES

The quantitative information and physical samples that undersea vehicles enable the nation's scientists and engineers to gather will play a central role in characterizing and protecting the Earth and the human environment; in developing food, fossil fuel, and marine mineral resources; and in enhancing national security and law enforcement.

The scientific challenge is to obtain new kinds of quantitative information about the Earth and its natural systems, sampled over a sufficiently broad range of space and time. This information includes physical, chemical, biological, and geological measurements. Undersea vehicles, with their evolving capabilities, can greatly improve the temporal and spatial resolution of many of these measurements. They can give scientists access to ocean environments not now accessible, such as under ice and at depths below 6,500 meters. Ocean scientists today use a combination of DSVs and ROVs to achieve these goals to a limited degree.

Commercial uses of underwater vehicles (as in support of offshore oil and gas production and fishing operations) are served by a competitive market for vehicles and vehicle services. The next few decades will witness expansion of commercial uses of the oceans. Oil and gas deposits on shore are being increasingly depleted. Fish and other renewable ocean resources require more intensive monitoring and regulation, as commercial fish stocks are more heavily exploited to feed growing world populations; sustainable development requires better knowledge of the fish populations and their ecological relations. Other uses requiring improved knowledge of the ocean environment include renewable energy and communications cable routes.

Commercial and scientific applications are mutually dependent. Fundamental scientific discoveries often lead to new and unexpected opportunities for commercial uses; examples include discoveries such as plate tectonics, high-temperature sea floor vents, methane hydrates, new living resources, and mineral deposits. In turn, scientific investigation is enabled by new vehicle capabilities pioneered in the commercial sector.

ROVs will remain the workhorse vehicles for commercial activities. But an increasing role is foreseen for AUVs, especially in survey, exploration, and environmental monitoring. New and improved sensors, guidance and control of task performance, navigation, data processing and storage, higher energy density, and improved reliability will be key technology development and improvement goals for long-duration AUVs. ROVs can be expected to be used in tasks of increasing complexity (such as changing chokes in undersea oil production flow lines) in response to industry demands.

Undersea vehicle programs for national defense did not grow as rapidly as was once forecast. However, demand by the military for such systems is likely to grow in the next decade. Reductions in defense spending will place a premium on less-expensive systems for underwater tasks involving labor-intensive operations such as surveying. Reductions in the number of military submarines will tend to increase demand for AUVs, which can perform some of the surveillance tasks commonly performed by submarines and can extend their surveillance range. Most undersea vehicles are transportable by aircraft for quick reaction to distant emergencies.

The U.S. Department of Defense has, and will continue to have, special information needs defined by military missions. An understanding of the physical environment, such as the factors affecting sonar performance in support of mine antisubmarine warfare or littoral topography and soil mechanics in support of amphibious assaults, is vital to the military. ROVs and AUVs may play expanding roles in obtaining such information by transporting sensors over a wide area from a single support ship. The military appears to be moving generally toward buying commercial equipment and using available technologies.

Law enforcement and regulation are government functions that hold important opportunities for undersea vehicles to contribute in the future. The USCG's missions of patrol and law enforcement include interdiction of drugs, pollution detection and control, and fisheries regulation enforcement. Innovative use of undersea vehicles could make the most of USCG assets by performing hull inspections for damage and contraband, searching for contraband on the seafloor, detecting and mapping pollution fields, and conducting reconnaissance and surveillance. Undersea vehicles will also help other law enforcement agencies, which often need to conduct accident investigations or recover bodies and evidence underwater. Many of the investigative needs of the USCG and other law enforcement and regulatory agencies can generally be met by commercial and defense-derived undersea systems.

SETTING PRIORITIES FOR UNDERSEA VEHICLE DEVELOPMENT

Summary of Potential Performance Improvements

Chapter 2 describes the current state of undersea vehicles and the potential for performance improvements as a result of technical advances across the range of subsystem technologies, including the following:

- DSVs of the present generation have reached a mature state of development. The next generation of smaller, lighter, and, perhaps, less complex DSVs will be based on advances in new materials, power, and pressure-tolerant electronics, when and if they are required.
- ROVs are widely used in undersea commercial activities and are increasingly used by the oceanographic scientific community. Incremental improvements in most of the subsystems of ROVs can be anticipated.
- AUVs require the most significant improvements in subsystems to enable their anticipated major contributions to national needs. Focused development in key subsystem areas should result in significant enhancement of overall system performance and capability.

Development Projects to Support Undersea Vehicle Advances

Identifying Development Priorities

Table 4-1 lists the key vehicle subsystems reviewed in Chapter 2 and summarizes the committee's evaluation of how improvements in each subsystem technology are likely to affect the overall performance of each of the three major classes of vehicle. The committee classified the subsystem technologies as "critical" (improvement would enable important new vehicle capabilities), "incremental" (improvement would benefit overall vehicle capability in an evolutionary sense), or "mature" (improvement would only marginally enhance vehicle performance).

The committee then ranked the critical subsystem technologies by priority: greatest potential (highest probability of benefiting from research to improve vehicle performance) and high potential (significant, but not quite as high, probability of benefiting from research to improve overall performance).

The committee considered many factors in constructing these rankings. The most important of these are (1) criticality of each subsystem's performance to the total system performance, (2) cogency of application to the three vehicle types (DSV, ROV, and AUV), (3) range of potential uses,

TABLE 4-1 Technology Assessment Summary by Subsystem

Subsystem	DSVs	ROVs	AUVs
Sensors	Incremental	Critical	Critical
Communications	Incremental	Mature	Critical
Mission and Task Performance Control	Mature	Incremental	Critical
Energy	Incremental	Mature	Critical
Navigation and Positioning	Incremental	Incremental	Critical
Data Processing	Incremental	Incremental	Critical
Signal Processing	Mature	Incremental	Incremental
Data Management and Storage	Incremental	Incremental	Critical
Work Subsystems and Sampling Devices	Incremental	Incremental	Critical
Launch/Recovery/Docking	Incremental	Incremental	Critical
Materials and Structures	Incremental[a]	Incremental[a]	Incremental[a]
Propulsion	Mature	Mature	Mature
Total System Integration	Mature	Mature	Critical

NOTE: Shaded areas indicate critical subsystems with greatest potential for vehicle performance improvement and total system integration.

[a]Critical for deepest applications.

and (4) degree of innovation required (which ranged from incremental improvement to radical innovation).

Other factors, such as agency mission priorities, are likely to influence the allocation of funds in practice. The committee's intent was to set broad priorities among the major vehicle subsystem technologies to provide general guidance for the allocation of research and development funds.

Results: Critical Subsystems with the Greatest Potential for Improving Vehicle Performance

The committee finds that three subsystem technology development areas offer the greatest potential for significantly improving the most-needed undersea vehicle data-gathering capability and task performance over the coming 5 to 10-year period: ocean sensors, undersea communications, and mission and task-performance control.

Ocean Sensors. Ocean sensors are needed in several areas that will enhance undersea vehicle performance:

Acoustic. Further development of a variety of acoustic sensors will improve undersea vehicle performance. These efforts should include sensors for more efficient and detailed seafloor and sub-seafloor mapping, assessment of plants and animals in the seafloor and in the water column, and tomographic mapping of the interior of the ocean. Development efforts should focus on the following:

- miniaturization (to permit use of more sensors per vehicle)
- reduction of power consumed in propulsion (to permit more on-board sensors and longer missions)
- combine different sensor types into single integrated systems (interdisciplinary advice—from biological, chemical, physical, and geological oceanography—is necessary for understanding data fields and processes)
- sensors for biological reconnaissance

Visual Imaging. Further development, miniaturization, and reduction of power consumption of optical, laser, and structured light sensors will allow existing capabilities to be integrated into smaller, less expensive vehicles. Improvements in resolution and data storage will allow closer matches between acoustical and optical surveys of underwater areas. On-board scene analysis will permit real-time remote control of AUVs at the task level, with tasks executed by the control subsystems.

Chemical. The development of chemical sensors is central to a number of scientific and environmental applications, including studies of gradient and plume phenomena. Continued progress will determine the usefulness of undersea vehicles in these areas. The field is developing rapidly, but more work needs to be done in several areas:

- In situ chemical analyzers capable of analyzing a variety of chemical species (including methane, organic contaminants such as polychlorinated biphenyls and hydrocarbons, helium, and specific radioisotopes), which cannot now be detected by technology appropriate to vehicles, must be developed.
- A method for accurate on-board long-term calibration of chemical sensors is needed.
- Sensors of low power and small size, which will permit incorporation of a variety of sensors on a single vehicle, must be developed.

Sensor Applications. Improved in situ sensing can greatly increase scientists' ability to determine the character and the temporal and spatial distributions of oceanic phenomena. Nearly 20 years ago, the CTD instrument provided greatly increased sampling density over reversing thermometers and Nansen bottles. The CTD data revealed important, previously unsuspected features. The committee believes that similar revelations are likely with regard to a variety of crucial oceanographic parameters, including nutrients, chemical species, hydrocarbons, and anthropogenic inputs. The prospect of integrating these multiple sensor modalities with arrays of fully mobile, intelligent vehicles provides a logical route to filling crucial gaps in the scientific understanding of the oceans from a multidisciplinary perspective.

Fisheries managers are among the potential beneficiaries. They must make decisions based on sparse data from a handful of stations, using techniques such as individual CTD casts and net tows. In the future, arrays could be used to improve forecasts based on simultaneous measurements of water temperature and salinity; nutrients; and the distribution of plant biomass, zooplankton, and fish on unprecedented scales of time and space.

Despite recent progress, many parameters, such as oil and many organic chemical pollutants, cannot today be measured in situ. Sensor development needs to proceed on a broad front so that new types of sensors, based on new technologies, can be incorporated into vehicles. Development of sensors will also be driven by the progress of scientific understanding of oceanographic phenomena. As long-duration undersea vehicles are developed, anti-fouling technology will be required to extend sensor usefulness and improve the reliability of long-term calibration.

Communications. The rapid advance of the technology of communications holds great promise. Fiber-optics provides a new medium for control and communications with tethered vehicles, with vastly greater bandwidth than conventional coaxial cables offer. Acoustic communication, on which DSVs and AUVs rely, has benefited substantially from the development of data compression techniques. A valuable and challenging research goal is to devise an acoustic modem with enough bandwidth to let human operators control a vehicle from a moving platform in reverberant and dynamic acoustic environments found in shallow water and at long horizontal ranges. Achieving that goal will allow human operators to remain in the vehicle control loop for some AUV operations. Communications between vehicles and shore via radio frequency/high frequency is another area of concern for certain missions.

Mission and Task-Performance Control. High-level mission logic and guidance for fault-tolerant AUV operation in a variety of applications in environments with wide levels of uncertainty would allow AUVs to become reliable tools and would encourage acceptance by users in industry, science, and the military. Further, development of sophisticated user-interface systems and techniques for control of ROVs and AUVs will improve human, high-level, real-time supervision of undersea vehicles and allow these vehicles to offer some of the in situ advantages of DSVs, including drawing directly upon human perception in near real time. The remote operator, for example, can use a light pen or mouse to augment the computer model of a televised scene and to instruct it. For example, the operator may designate a specific feature of the televised scene (such as an isolated organism) for the vehicle or camera to follow automatically.

Task-performance control encompasses the operation of work systems and physical sampling devices, which is discussed later in this chapter. As an integral part of the total human-vehicle-manipulator control system, the vehicle and its manipulators or sampling devices can move in concert to carry out tasks specified graphically at the object level through a graphical user interface. This coordination can be done in real time, even by an AUV, due to the much smaller communication bandwidth required by high-level instruction, combined with improvements in acoustical bandwidth now becoming available. These efforts should include fault-tolerant mission logic, layered control architecture, virtual-presence operator interfaces, and object-based, task-level control.

Results: Other Critical Subsystems with High Potential for Improving Vehicle Performance

The committee has also determined that there are other critical areas of ocean-unique technology with high potential for improving undersea vehicle performance.

Energy. Continued development of high-performance energy storage technologies will expand the scope of undersea vehicle applications. Current energy technology, however, is adequate for AUVs to be applied in some applications. Technology programs applicable to this challenge can be found in the automobile industry's electric vehicle programs (which are assisted by the federal government), and in the programs of other nations, such as Canada's aluminum-oxygen battery development program.

Navigation and Positioning. Integrating inputs from a variety of navigation sensors can create navigation systems that are repeatable and can operate independently of emplaced seafloor references. These developments will improve the performance of AUVs in particular, especially for applications requiring missions of long durations and distances. This integration should include inertial packages, bottom-lock sonars, and bottom-lock optical and video subsystems.

Data Processing. Greater data processing capability will be required by the vastly increased on-board data production of new sensor suites and control systems. This capability will permit real-time interaction with the data to provide an ongoing quality control during the vehicle's mission. Data processing algorithms for the automatic recognition of targets and features in real time will bring high-level inputs into AUV mission logic, enabling the vehicle to react better to its environment and reducing data storage requirements.

Signal Processing. Related to data processing, on-board signal processing will be particularly important for AUVs, because navigation and target detection will be critical to mission performance.

Data Management and Storage. New techniques for storing massive amounts of data (up to 1 gigabit) on chips as small as 16 cubic centimeters (1 cubic inch) need to be adapted to undersea vehicles to greatly increase on-board data storage capacity. The accompanying reduction in the space required for storage and lower energy consumption are additional benefits. In addition to on-board storage, systems need to be developed for uploading data during missions, perhaps by burst transmission to satellites or by physically plugging into a data transfer network on the seafloor. For longer duration missions such systems would enable the vehicle to regain previously used storage capacity after each transfer.

Work Systems and Physical Sampling Devices. Work systems include all external (i.e., bolt-on) on-board devices that permit doing physical tasks at the work site. They include manipulators, tool packages, and "end effectors," such as cutters, drills, and probes. Physical sampling is the removal of living and nonliving materials from the ocean for study at a remote site. While the development of sensors and on-board in situ analytical devices will greatly reduce the need to take some types of physical samples, there will still be a need for specialized sampling devices. Some requirements for these devices are as follows:

- external and internal water samplers, free of all trace element contaminants, that are gas tight (capable of holding pressure even if samples are supersaturated with gas)
- devices for capturing and maintaining live marine organisms at ambient temperatures and pressures
- tools for boring, coring, and cutting rock samples and methods for uncontaminated storage of samples
- devices for collecting seafloor sediment samples and maintaining them uncontaminated at in situ pressure

Materials and Structures. For the deepest ocean applications, new materials and design strategies will enable new capabilities and applications for all three vehicle types. These efforts should include (1) glass ceramic pressure hull materials for extreme depths, (2) lower cost long-filament fiber-optic links, and (3) marine biofouling coatings for long-duration submersibles.

Launch, Recovery, and Docking. Virtually any undersea vehicle operation requires a launch and recovery system provided by a mother vessel. The system is basically an elevator, often including a cage housing the vehicle. Vehicle operations are generally limited by the sea-state conditions under which launch and recovery can be conducted. Operational delays due to high sea states are costly and can considerably degrade productivity at the dive site. Progress has been made in these systems, but improvements are still needed, particularly for handling ROV tethers.

Total System Integration. The goal of the system integration process is to bring together subsystems technologies to prove the system's level of technical maturity, assess its potential for commercial or scientific application, and establish a basis for cost-effective design—all in the context of a specific mission or missions.

During the last two decades of government-sponsored development, much emphasis has been placed on the subsystem or component level. Where advanced concepts and technologies have been integrated into a system, the objective has generally been to respond to achieving a specific Navy application or mission; cost objectives have not been a high-priority, and potential commercial uses have not been a consideration. System integration emphasizes the process of transferring technology to mature systems—a process involving elements, such as safety, handling, personnel support, and cost performance—as well as vehicle design concerns, such as the use of modular subsystems and payload integration. Integration will not develop new technologies, but it is the step to commercial and cost-effective use that is most often avoided by those engaged in development.

Many of the vehicle system integration activities that have occurred have been classified, because they have been mission specific. In the committee's view, the result is that engineers and scientists in industry and academia have become aware of advances in technology, but they have seldom been exposed to the problems and successes of the integration process or the costs of the system and subsystems and their operation.

Since the end of the Cold War, with reduced military budgets and increased openness about sharing in development, there have been attempts to foster dual-use development. Advanced technology demonstrations have focused on systems integration and making technologies available to the private sector users. However, military specification requirements still prevail and severely constrain the cost-effectiveness of vehicle development. Transfer of government-sponsored, vehicle-related development to the private sector for wide-range, reasonably priced, applications must entail a balanced development program of technology advancement and systems integration. For example, fuel cells, advanced battery chemistries, closed-cycle internal and external combustion engines, and numerous other concepts have been funded at the technology level, but few of these concepts have moved to the private sector because of lack of integration and the resulting high cost.

Table 4-1 summarizes the probable impact upon vehicle performance of technical developments in each of these subsystems. The table suggests that AUVs offer the greatest potential improvements in return for investments in development. In every case, it is necessary to be sensitive to the need for total system design and integration.

FINDINGS

Finding. Undersea vehicles present an unusual confluence of technical opportunities with national needs. The human needs for data-gathering and work under the oceans are growing steadily. The technologies of information processing, microelectronics, and communications are evolving just as quickly, offering improved vehicle capabilities. With these capabilities, oceanographic, hydrographic, and environmental information could be obtained at lower cost, with higher accuracy and reliability. The development and integration of the technologies involved are relatively straightforward.

Finding. The committee has found that the subsystem technologies with the greatest potential for significant improvements in performance are sensors (acoustic, visual, and chemical); communications (both acoustic and digital fiber-optic techniques); and advanced guidance and control methods that then enable task-level direction of vehicles and even fully automated operation.

Finding. These advances will enhance all classes of vehicles to some degree. AUVs, however, are likely to benefit most from further investments in development. AUVs are emerging from the research stage and have enormous potential for a wide variety of missions in support of science, industry, and government. The main development goals are improved sensors, guidance and control, navigation, and data processing, as well as higher reliability. Rapid advances in the technology of data processing and control raise the important possibility of hybrid AUVs, with communication links (either acoustic or fiber-optic) that carry control signals. The main burden of developing AUVs, perhaps for the next 5 to 10 years, will fall on government; the technology is not commercially mature. For that reason, and because many of the AUV applications are highly specialized and the vehicle technologies must be developed or adapted for undersea use, the development and integration of most AUV systems are too risky for industry and users to carry forward without help.

Finding. ROVs are likely to remain the "workhorses" of commercial activities. They will see increasingly wide applications, paralleling the increasing human dependence on the ocean and its resources. These applications will include offshore oil and gas development and support of operations, telecommunications cable work, and search and recovery missions. ROVs will also serve defense, law enforcement, and regulatory missions, such as underwater surveillance, fisheries monitoring, and pollution detection. Some oceanographic and hydrographic studies will also be carried out using ROVs. The manufacturers and users of these vehicles can be relied on to develop the technology and to adapt to special noncommercial needs.

Finding. The technology of DSVs is generally mature. Improvements in performance and cost will come with advances in materials, power systems, and electronics. The market for these one-of-a-kind systems is small (mainly government) and will be served by specialized engineering firms. The current fleet of DSVs, although aging, will continue to serve military and scientific needs.

Finding. The U.S. Department of Defense, and in particular the U.S. Navy, has perhaps the broadest requirements for undersea vehicles for a variety of missions, including search and retrieval, surveillance, mine detection, and exploration of the operational environment (such as sonar performance and coastal topography). ROVs, including those from commercial sources, and AUVs will become increasingly important additions to the military fleet and will justify a government-funded program of development and technology integration. Regulatory and law enforcement agency requirements can be met through the adaptation of commercial and defense vehicle systems.

The committee has devoted much attention in this report to individual subsystem technologies, assessing their near-term potential contributions in practical vehicles. The reader should not be misled by this focus on subsystems. Vehicles consist of integrated systems. The challenge of system integration is more difficult, and surely more important, than technology development at the subsystem level.

5

Enhancing the Nation's Capacity for Undersea Work and Research

To take best advantage of the capabilities for meeting national needs, such as those identified in Chapter 3, the nation will need an orderly process for identifying requirements and making available the systems to meet them. This committee does not favor a single overarching policy governing development and use of undersea vehicles, however. The vehicles, their missions, and their users are too complex and varied for such a policy. Technology development, for example, is carried out in this country by various organizations—public and private, civilian and military—and in other countries, as well. Users are a similarly diverse lot, motivated by a tremendous variety of scientific, industrial, and governmental interests (summarized in Chapter 3). These organizations have evolved their roles and relationships over the years in efforts to solve short-term problems. A process of strategic planning, by which consensus can be reached on major long-term goals, can ensure that the needed resources will be available without degrading the healthy diversity and pluralism of the system.

In Chapter 2, the committee reviewed the opportunities presented by the evolving technology of undersea vehicle systems; national needs to which these vehicles could be applied were identified in Chapter 3. In Chapter 4, the committee built on these two assessments, proposing development priorities for each subsystem technology and each vehicle type, to improve the nation's ability to carry out missions addressing those needs.

The nation's undersea vehicle capabilities have three vital components:

- technology development and technology transfer by the private and public sectors to ensure that vehicles are capable of performing the missions required
- capital investment necessary to provide new vehicles and upgrade existing ones to meet identified national needs
- availability of vehicles to scientific users, with effective scheduling and coordinating services, efficient and capable support vessels, and adequate operating funds

This chapter reviews current trends with respect to each of these components. It offers alternatives for the public and private sectors to address deficiencies by setting and pursuing long-term goals.

NEED FOR A STRATEGIC APPROACH

The scale of the Earth's oceans and their resources, and the complexity of the institutions for working in them and exploring them, argue strongly for a coherent strategic plan. Lacking such a plan, the agencies involved have no disciplined process for agreeing on their long-range goals. Instead, they have evolved effective but informal networks for the ad hoc discussion of problems in the short run. However, for federal applications of undersea vehicles, including science, various salvage tasks, and national security missions, the alternative to planning is inevitably a decline in capability. The market will provide for private sector needs.

Mismatched Goals

An illustration of the mismatch among agency goals can be found in the variety of organizations that provide vehicles for research and work in the deep ocean. Three federal agencies—the National Science Foundation (NSF), NOAA's National Undersea Research Program (NURP), and ONR—evaluate and fund proposals for most scientific uses of undersea vehicles. The research funding agencies also provide some funds for developing sensors and other vehicle subsystems, as does DARPA.

Except for *Alvin* and the ROV *Medea-Jason*, which are U.S. Navy-owned but available full-time for science, most of the deep ocean vehicles are owned and operated by the Navy and made available part-time for scientific purposes, subject to the unpredictable operational requirements of the Navy. For reasons of cost and scheduling convenience, the Navy-operated vehicles do not always match the requirements of scientific users. However, they are being increasingly used by researchers, owing to improved, but still informal, coordination between the U.S. Navy's Deep Submergence Office and NSF, NURP, and ONR (Dieter, 1996).

Other independent centers for undersea research and technology, private and public, have sprung up over the years, with the support of NURP, NSF, the U.S. Navy's Supervisor of Salvage, and others. The most notable of these include:

- the Hawaii Undersea Research Laboratory, one of six NURP regional centers, which operates the 1,500-meter depth capability DSV *Pisces V*
- the private, nonprofit Harbor Branch Oceanographic Institution, which operates the DSVs *Sea-Link 1* and *Sea-Link 2* (both with 800-meters depth range) and the 350-meters DSV *Clelia*, for lease to research funding agencies.
- the private, nonprofit Monterey Bay Aquarium Research Institute, which operates the 1,850-meters ROV *Ventana*, with a 4,500-meters ROV, *Tiburon*, in development (Fox, 1994)

Thus, the research funding agencies have a variety of sources of vehicles, with different financial practices, scheduling systems, and constraints on availability. Because the organizations that use undersea vehicles are not those through which they are funded, maintaining adequate support is difficult, particularly in times of budget cuts. The personnel of all these organizations work together to accommodate each other's needs, but without formal agreement on goals they cannot effectively bring their priorities into harmony.

Finally, users of the national undersea capability must acknowledge the leadership of foreign programs, which have invested heavily in advanced undersea vehicle and subsystem technologies (Appendix B). The leader in deep submergence technology today is the well-funded Japanese program, which has the high-priority support of government and the core mission of helping understand seismic activity in nearby ocean trenches. Some U.S. researchers have used the deep-diving Japanese vehicles, under the auspices of National Undersea Research Program. NOAA has engaged in scientific information exchanges with France, through the U.S.-French Cooperative Program in Oceanography. NOAA also has a cooperative research agreement with JAMSTEC, through the U.S.-Japan Natural Resources Cooperative Program. This arrangement has given U.S. users access to the deep-diving Japanese vehicles. In AUV technology, Russia is probably the leader, with an array of operational deep-submergence AUVs (Mooney et al., 1996). International cooperation in both technology and research are vital.

Planning without Central Control

A plan that took all of these varied assets into account would need to look ahead at least a decade, with goals based on high-priority national needs, such as those reviewed in Chapter 3. No such national needs have been enunciated as yet; this committee, lacking a statement of national policy, has not offered recommendations on that score. The plan would link those needs with specific vehicle capabilities and survey the public and private technology and capital investment programs, here and abroad, to ensure that they were being met.

The plan need not lead to a strictly coordinated program of funding for undersea research. A comprehensive national approach can build on the pluralism of today's varied programs, allowing for healthy short-term flexibility and individual initiative. At the same time, a consensus on future needs, embodied in a strategic plan, would provide guidance to the agencies responsible for making the necessary investments in technology and in new vehicles. The plan would afford U.S. scientific users, for example, the widest possible choice of undersea systems, both here and abroad, and allow the U.S. Navy and the research funding agencies to cooperate more fully on the next generations of vehicles. The planning body would act as a board of directors for all missions in the federal sphere. As such, it would need to include representatives from users, developers, and operators, both inside and outside of government. Ocean and Earth scientists, for example, would be prominently represented, as would the national security interests of the U.S. Navy, which is responsible for building and maintaining most of the vehicles used for scientific research; the Supervisor of Salvage; and liaison representatives of foreign undersea vehicle programs.

The planning body would have secure support from a single federal agency and would be given statutory authority and a small, but sufficient, administrative budget. The body would be limited in function to planning, with no technology or research programs or capital assets, to avoid conflicts of interest with the various mission agencies it serves.

Two existing organizations offer models that could be built on to establish such a body:

- **Joint Oceanographic Institutions, Inc. (JOI).** JOI is a private, nonprofit consortium of 10 U.S. academic institutions with the mission of managing and planning research programs in the ocean sciences. Among its tasks is carrying out U.S. elements of the international Ocean Drilling Program and the Nansen Arctic Drilling Program; coordinating the global Ocean Seismic Network; and promoting the integration and networking of computers and communications in ocean research. JOI's consortium for oceanographic research and education (CORE) promotes partnerships of government, industry, and academia. The international liaison function of JOI is of particular interest.
- **Deep Submergence Science Committee (DESSC).** The nucleus of a fully governmental planning function for undersea vehicles is available in DESSC, a joint program of NOAA, NSF, and the U.S. Navy, under the umbrella of the University-National Oceanographic

Laboratory System (UNOLS). DESSC promotes the most effective use of undersea vehicles, mainly through scheduling scientific use. DESSC oversees the use of *Alvin* and the U.S. Navy-owned ROV *Medea-Jason* and helps review proposals for academic use of the Navy's other deep submergence assets under a Navy-NOAA agreement (UNOLS DESSC, 1993). The committee also advises research funding agencies (NSF, NOAA/NURP, and ONR) on the state of vehicle technology and applications. But today DESSC is restricted to scheduling research use of the Navy's deep submergence vehicles (*Alvin* and *Medea-Jason* at Woods Hole Oceanographic Institution and the two Navy-operated DSVs, *Turtle*, and *Sea Cliff*). Beyond its own self-initiated studies (e.g., Fox, 1994), DESSC has no long-term planning role. In addition, its charter is too narrow—covering only a fraction of research activities and with no voice in decisions bearing on technology development or capital investment.

The committee has not assessed the relative strengths and weaknesses of these models. It simply stresses the need for a competent planning body.

TRENDS AND POLICY ALTERNATIVES

Inefficiencies can be found in each of the three aspects of the national capability considered by the committee (technology development, capital investment, and availability and access). There is a lack of coordination between user and operator agencies, and they lack a strategic approach to their requirements. As a result, too little federal investment in technology, inadequate capital investment (with aging and increasingly obsolete facilities), and undue constraint on operating funds. The committee has identified policy options for addressing these deficiencies in the context of strategic planning.

Technology Development

Chapters 2 and 3 reviewed the key technology issues for the various types of undersea vehicles. The highest priority technologies, in general, are in three areas: sensors, communications, and control. Policy makers should recognize that advanced undersea vehicle and subsystem technology programs can be found in a number of nations throughout the world and that the United States should be alert to opportunities to share the costs and benefits of its technology investments internationally. The oceans, after all, are global in scope, and the scientific and industrial problems of their waters and the seabed are likely to be of wide and abiding interest.

Deep Submersible Vehicles

With few exceptions, the technology of human-occupied submersibles has been developed by governments or by industries with government support. In the United States, the U.S. Navy has assumed this role as an offshoot of the national security applications of DSVs. Declines in Navy spending, however, have left the DSV fleet of the United States obsolete in many ways. *Alvin* (operated by the Woods Hole Oceanographic Institute, 4,500-meter depth range) and *Turtle* (operated by the Navy, 3,000 meters), are the mainstays of the nation's deep ocean research. They were built in 1964 and 1968, respectively (and last modified in 1974 and 1984, respectively). The 6,000-meter *Sea Cliff*, operated by the U.S. Navy and also available part-time for research purposes, was built in 1968 and modified in 1985. The two Johnson *Sea-Links* and *Clelia* (developed and owned by the private Harbor Branch Oceanographic Institution) date from the mid-1970s (see Appendix E).

While the Navy has the main responsibility for developing DSV technology, the associated technologies of sensors and some other payload systems have been advanced to some degree by agencies oriented toward academic users, notably NSF, NURP, and ONR.

Many other nations have strong DSV technology programs (Appendix B). The well-funded and ambitious undersea vehicle program of JAMSTEC operates a 6,500-meter DSV, *Shinkai 6500*, in its undersea research. *Shinkai 6500* is the deepest diving DSV in the world today and, with further development, is expected to go deeper still, to the very bottom of the deep ocean trenches near Japan. Russia and the Ukraine have sizable, capable fleets of deep-diving DSVs, including the Finnish-built *MIR* class (with a depth range of 6,000 meters), and have developed new techniques for fabricating high-strength hull materials. The French DSV *Nautile* also has a 6,000-meter depth range.

Remotely Operated Vehicles

ROV technology was first developed by the U.S. Navy in the 1960s, but it has been advanced mainly by the private sector since the 1970s. The offshore oil and gas industry and its service organizations led the way, followed by the cable-laying operations of the telecommunications industry. ROV technologies are now mature; further development will be driven by the market demands of those industries and others, notably the salvage and treasure-hunting industries. Some commercial ROVs have been adapted to carry out scientific missions, with specialized sensor suites: a notable example is the ROV *Ventana*, operated by the private, nonprofit Monterey Bay Aquarium Research Institute (Newman and Robison, 1993).

An important exception to the private-sector ROV rule is the technology of deep-diving ROVs for research. In the late 1970s the U.S. Navy developed an ROV called the *Advanced Tethered Vehicle* (*ATV*), with a depth range of about 6,000 meters. This vehicle, developed for national security purposes, is used for scientific missions in conjunction with the DSV *Turtle*. The *ATV* was recently used in NOAA studies of the petrology and hydrothermal process at the Blanco Transform Fault Zone.

In Japan, JAMSTEC has built a deep-diving ROV, *Kaiko*, with which it reached the bottom of the deepest point in the world's oceans, in the Mariana Trench, in early 1995.

Autonomous Undersea Vehicles

AUVs are still largely developmental systems, although they have seen some limited operational use in national security and scientific missions. The U.S. Navy's 6,000-meters *AUSS*, is operational and available for scientific missions, but is generally considered too big and costly for such uses (Fox, 1994). As noted in Chapter 4, AUVs have greater potential than other vehicle types for benefiting from research and development. The U.S. Navy, NURP, NSF, and DARPA have all sponsored the development of AUVs and AUV subsystems. NSF, for example, spends about $700,000 annually on the developmental vehicle autonomous benthic explorer (*ABE*), which will be used for unattended missions of up to a year (Clark, 1996).

Development of AUV technologies has benefited from a vigorous informal network of engineers and scientists representing oceanographic organizations, government agencies, and contractors for the U.S. and Canadian governments. In the next decade or so, with foreseeable advances in sensors, communications, and control, and energy supplies, as well as more efficient energy use, these vehicles will become fully capable scientific research vehicles. Because further development of AUV technology depends on public investment, the expected reductions in government research and development funding are likely to slow AUV development.

Some foreign AUV programs are quite successful (Appendix B). The Russian Institute for Oceanological Problems in Vladivostok, established in 1988, with a scientific staff of about 90, has developed a small fleet of highly capable AUVs (depth ranges down to 6,000 meters) and used them in deep ocean search and recovery operations. These AUVs have made several hundred operational dives deeper than 4,200 meters (Mooney et al., 1996). Both the French and Russian programs have 6,000-meter AUVs. The United Kingdom National Environment Research Council, with aid from the European Union, is spending £700,000 (about $1 million) per year on designs for an AUV to be known as "Autosub." It is intended to cross the Atlantic Ocean, diving to the bottom for periodic core samples and other measurements, then surfacing to transmit data via satellite (Seymour et al., 1994). The system is expected to be operational by 1998 (Griffiths, 1996).

Policy Options and Technology Transfer

In technology development the nation can meet its needs either by its own efforts or by cooperating or entering into partnerships with others. The goals must be clear, however, with guidance from the strategic planning body described in this chapter. Three approaches can be taken, separately or in combination, to improve the technology of undersea vehicles:

- Continue to maintain or enhance national development of technology through the U.S. Navy, NSF, NOAA, and DARPA, emphasizing technologies that are critical to improving of vehicle capabilities but not likely to be developed in response to commercial demand. Wait for foreign development of the technology and invest in adaptations of that technology.
- Establish joint technology development agreements with foreign programs, such as those in Japan, Russia, and France (Appendix B).
- Develop cooperative industry-government programs to improve technology development and transfer.

Complete reliance on the first option is not realistic in today's federal budget circumstances. In any case, it would be an inefficient use of resources to rely solely on government-developed technology, given the wealth of foreign technology and the opportunities for technology transfer from industry in critical subsystem technologies such as energy supplies, sensors, navigation, and control. Guided by the goals of an agreed-upon strategic plan, U.S. technology developers and users can identify needs and the technologies to meet them, working jointly or singly, as circumstances dictate.

Capital Investment

If the United States is to retain the ability to do meaningful work under the sea, it must pay constant attention to the capital assets embodied in both undersea vehicles and their supporting systems. These assets today are a highly capable and varied fleet of DSVs, ROVs, and AUVs. They are supplemented by a wealth of foreign systems, including the deepest-diving DSVs and ROVs in the world at JAMSTEC. Strategic planning must confront issues of priorities in identifying which systems to retain as national assets, which to invest in, which to mothball, and which to decommission.

The U.S. fleet of DSVs is aging. *Alvin* dates from 1964. Although it was rebuilt with a new titanium hull in 1973, extending its depth rating to 3,658 meters, its design still places fundamental limits on payload, power, maneuverability, and observational capabilities (Fox, 1994). The youngest U.S. Navy-owned vehicles, *Turtle* and *Sea Cliff*, were launched in 1968. The Hawaii-based *Pisces V*, built in 1973, is limited in depth to 2,000 meters. The committee found no compelling scientific rationale for federal investment in a new DSV system for use in the deepest ocean, especially in view of the active Japanese programs, which welcome participation by scientists and engineers from the United States and elsewhere. As time goes on, results from Japan and elsewhere could provide such a justification.

Most new scientific ROVs are adaptations of commercial systems, which are freely available on the open market. These systems, and the few more specialized government-developed vehicles, such as *ATV* and *Medea-Jason*, are seeing increasing use in scientific research as the technologies of sensors, communications, and control improve. In general, private industry can be relied on to continue providing ROV systems.

The federal government, on the other hand, is responsible for providing AUVs to users. The U.S. Navy's large, complex *AUSS* is the only operational AUV in service. NSF's *ABE* is nearing operational status, having undergone sea trials for the past few years. The small developmental AUV *Odyssey* saw its first operational tests in 1995. The potential of AUVs is very important, but the systems are not yet mature enough to be of interest to industry, so further federal investments will be necessary.

Policy Options

There are two alternatives for bringing capabilities into line with future needs:

- Develop or buy the necessary systems and supporting vessels and provide them with the necessary operating funds.
- Meet as many needs as possible through partnerships with foreign programs and with the private sector. These partnerships might range from simple leases of U.S.-certifiable foreign submersibles to joint development and use of new vehicles and support vessels. It should be noted that most foreign DSVs are not U.S.-certifiable.

It is likely that both measures will be needed, particularly in today's strained budget circumstances. The large investments necessary for undersea vehicles cry out for sharing costs and benefits as widely as possible.

Access to Undersea Vehicles

For the commercial uses to which undersea vehicles are put, the market provides access to the desired systems. For purposes in the public interest, such as science and national security, various federal agencies have assumed responsibility for making available the necessary vehicles and support systems. Again, the need for a strategic approach is conspicuous.

Several federal agencies and private foundations share the responsibility for providing undersea vehicles for research. Most prominent are the DSV *Alvin* and the ROV *Medea-Jason*, owned by the U.S. Navy but operated by a joint venture of NSF, NURP, and ONR at Woods Hole Oceanographic Institution; the DSVs *Turtle* and *Sea Cliff* and the ROV *ATV*, operated by the Navy's Submarine Development Group One; the DSV *Pisces V*, operated by NURP's Hawaii Undersea Research Laboratory; and several DSVs operated by the Harbor Branch Oceanographic Institution for lease to researchers and others. ROVs for research in relatively shallow water are owned and operated by universities, including NURP centers, and private foundations.

Academic research using undersea vehicles is funded by NSF, NURP, and ONR through standard academic peer review arrangements. U.S. Navy-owned vehicles are available to the funding agencies for the cost of fuel and other consumable vehicle supplies; others are leased at full cost, without subsidy. The DESSC of the University-National Oceanographic Laboratory System, a joint venture of NSF, NURP, and ONR, provides scientific coordination for *Alvin* and *Medea-Jason* and, through a 1993 agreement with the Navy's Deep Submergence Office, assists with scheduling scientific use of *Turtle* and *Sea Cliff*, *NR-1*, and *ATV*. Coordination for other vehicles is done by the particular funding agencies.

Funds for operating and maintaining the Navy's vehicles are under strong budget pressure, illustrating the mismatch between users and funders. The U.S. Navy Deep Submergence Office budget for engineering support has declined by more than 55 percent since 1987, and its vehicle operating funds by more than 20 percent (Lancaster, 1996). These cuts suggest that the parent agency, the Submarine Force, has placed a higher priority on building and maintaining its attack and strategic submarines. The Submarine Force derives little benefit from the undersea vehicles discussed in this report, although it funds them. Loss of Navy support of these vehicles, which could be considered national assets (essential to the national interest in national security, ocean research, or search and salvage), would dramatically increase the unit costs of the research to funding agencies because the Navy's operating cost subsidy would no longer be available. It would also raise costs of other users, such as the Navy's air forces and the U.S. Navy's Supervisor of Salvage. Maintaining the appropriate national asset facilities would be easier with a long-term plan, through which agencies and users could establish requirements based on agreed-upon national needs. A more stable funding scheme for undersea vehicles that are determined to be national assets needs to be developed. ONR's contributions to the National Deep Submergence Facility at Woods Hole Oceanographic Institution have decreased substantially, too, but NSF funds have risen to compensate, so the overall recent trend is steady, although uncertain (Dieter, 1996).

Research funding is also declining. NURP's budget, for example, has fallen from an all-time high of $18.1 in 1994 to $14.4 million in 1995, and only $12 million under the 1996 congressional continuing budget resolution (Kalvaitis, 1996). NURP's very existence is in doubt; only last-minute congressional action prevented its shutdown in fiscal year 1995, and similar rescues occurred in several previous years.

Coordination of these assets is obviously growing more important. Caught between declining military budgets and expanding scientific frontiers, the community must make better use of all its resources.

Policy Options

The committee believes that there are several alternatives for coordinating access—from the joint strategic planning described in this chapter to full consolidation of undersea vehicle support in a single federal agency. Options include:

- establishing an interagency coordination body for strategic planning, without administrative responsibilities
- consolidating support for the scientific use of undersea vehicles in a single agency, with a consistent approach to users and a capability for strategic planning with an eye to the funds available for investment in capital assets and technology development
- establishing leadership within a single agency to ensure not only access to vehicles, but also the necessary capital investments and technology development, in accord with a long-range plan that balances asset development, support requirements, research requirements, and available funding. (An approach much like this, but restricted to deep ocean vehicles, has been proposed by UNOLS DESSC [Fox, 1994]).

These three approaches are arranged in ascending order of centralization and control. Central control has the advantages of certainty and simplicity, but it entails a sacrifice of flexibility and local innovation, which are major strengths of the U.S. system of science and technology. A central authority would inevitably assume the problem of adjudicating among conflicting interests in different areas of research (for example, fisheries and geochemistry) and among agencies and institutions. The priorities of existing agencies should be respected. At the same time, a shared strategic view of future needs—developed through a disciplined planning mechanism—would be more effective in making the desired vehicles and support services available.

FINDINGS

Finding. Technology development, capital investment, and access to vehicles are the three components of a well-balanced national capability to work under the sea. An effort to coordinate the scientific and public interest applications of undersea vehicles must take account of all three.

Finding. The government-operated systems for undersea science and salvage are not optimally supported, especially for science.

Finding. Future requirements for such systems have not been given systematic attention. A strategic plan to meet national needs would identify appropriate roles for the private and public sectors in meeting those needs.

REFERENCES

Clark, L. 1996. Personal communication to D.M. Brown, March 20, 1996.

Dieter, E.R. 1996. Personal communication to D.M. Brown, March 20, 1996.

Fox, P.J. 1994. The global abyss: The need for an integrated national program. Marine Technology Society Journal 28(4):60–67.

Griffiths, G. 1996. Personal communication to Donald W. Perkins, April 23, 1996.

Kalvaitis, A. 1996. Personal communication to D.M. Brown, March 25, 1996.

Lancaster, E.L. 1996. Personal communication to Donald W. Perkins, March 25, 1996.

Mooney, J.B., H. Ali, R. Blidberg, M.J. DeHaemer, L.L. Gentry, J. Moniz, and D. Walsh. 1996. World Technology Evaluation Center Program. World Technology Evaluation Center Panel Report on Submersibles and Marine Technologies in Russia's Far East and Siberia, in press. Baltimore, Maryland: Loyola College of Maryland, International Technology Research Institute.

Newman, J.B., and B.H. Robison. 1993. Development of a dedicated ROV for ocean science. Marine Technology Society Journal 26(4):46-53.

Seymour, R.J., D.R. Blidberg, C.P. Brancart, L.L. Gentry, A.N. Kalvaitis, M.L. Lee, J.B. Mooney, and D. Walsh. 1994. World Technology Evaluation Center Program. Pp. 150–262 in World Technology Evaluation Center Panel Report on Research Submersibles and Undersea Technologies. NTIS Report No. PB94-184843. Baltimore, Maryland: Loyola College of Maryland.

University-National Oceanographic Laboratory System Deep Submergence Science Committee (UNOLS DESSC). 1993. Terms of Reference (revised July 13, 1993). Narragansett, Rhode Island: University of Rhode Island.

6

Conclusions and Recommendations

CONCLUSIONS

On the basis of its review, the committee developed the following conclusions with regard to undersea vehicles.

Conclusion 1. The nation has vital economic and scientific needs to significantly advance its capabilities for working, monitoring, and measuring in the ocean. Those needs involve national security, environmental protection, resource exploitation, and science. Undersea vehicles can contribute strongly to these capabilities by giving human beings access to new kinds of information about little known areas of the ocean and the seabed—information that may have a major impact on the well-being of large populations.

Conclusion 2. Technical advances are needed if the nation is to realize this potential; the priorities for these technologies are ranked in Chapter 4. The nation needs the ability to carry out construction support tasks, inspection, and maintenance in deep sea oil and gas fields safely and efficiently using remote control. Autonomous undersea vehicles in synchrony with research vessels can help gather oceanographic and hydrographic data more accurately, quickly, and cheaply. Monitoring pollution and measuring the conditions that could lead to global climate changes will be easier with new chemical sensors. Surveying the bottom with the high resolution offered by undersea vehicles is likely to reveal valuable mineral deposits and assist in the location and recovery of objects related to public safety and security.

Conclusion 3. The committee finds the technological advances most critical to these important missions are in the areas of ocean sensors, subsea communications, and mission and task-performance control systems.

Higher energy-density power sources are also important but very costly. The undersea vehicle industry must rely heavily on research and development in the automobile, aerospace, mobile communications, and microcomputer industries, which have made large investments in this field of technical development.

Conclusion 4. The vehicle technologies are generally mature enough to place the emphasis of technology advancement programs appropriately on systems integration. These mature technologies will permit systems like the following to be built:

- vehicles with endurance that can spend weeks on the bottom and are able to cover large areas of the bottom with high-resolution surveys, with sensor modules suited for varied missions
- AUVs designed to work in parallel with survey vessels, increasing the efficiency and accuracy of oceanographic sections and hydrographic surveys
- AUVs able to operate either under high-level human command or independent of human control for periods of months on data-gathering missions to provide detailed information about ocean dynamics, including physical, biological, and chemical processes

Conclusion 5. Other countries today, like the United States in the past, have mounted focused programs with sustained support in the service of well-defined national needs.

Conclusion 6. The United States has no concerted program; instead it has a number of informally coordinated programs and no disciplined mechanism for long-term planning. The financial disjunction between users and the federal providers of undersea vehicles in some cases impedes coordination.

Conclusion 7. Failure to address the deficiencies of federal programs will constrain scientific progress, limit the nation's ability to develop and manage its ocean resources, and compromise national security and law enforcement.

RECOMMENDATIONS

Recommendation 1. The nation should develop, maintain, and follow a long-term plan for federal undersea vehicle capabilities that takes into account all of the available facilities for undersea research.

CONCLUSIONS AND RECOMMENDATIONS

- The goals set out in the plan should be based on a still-to-be defined set of high-priority national needs in research and other federal missions. The specific capabilities of vehicles (deemed "national assets") should be measured against national needs to ensure that they are being met by public or private programs here and abroad. (The market will meet the needs of the private sector.)
- The plan need not lead to a strictly coordinated program of funding for undersea research. The pluralism of today's varied programs allows for healthy short-term flexibility and individual initiative.
- The plan should include provisions for a variety of projects, like those outlined in the focal projects (Chapter 3) based on agreed-upon national needs.
- The planning body should include representatives of users, developers, and operators, inside and outside of government, including ocean and Earth scientists, national security officials of the U.S. Navy (which is responsible for building and maintaining most of the vehicles used for scientific research), the Navy Supervisor of Salvage, and liaison representatives of foreign undersea vehicle programs.
- The planning body should have secure support from a single federal agency and should be limited in function to planning.

Recommendation 2. In developing undersea technology the nation should meet its needs through combining government programs, joint technology agreements with foreign programs, and cooperative industry-government programs. Maximum use should be made of programs outside the federal government. All decisions should be based on the long-term plan recommended in this report.

Recommendation 3. Capital investment programs should take advantage of partnerships—from leases of U.S.-certifiable foreign submersibles to joint development and use of new vehicles and support vessels with industry and foreign programs. The federal government should buy wholly new vehicles for civilian use only when other sources are not available and the national interest (as determined by the planning process recommended here) demands it.

Recommendation 4. In ensuring user access to undersea vehicles, the nation should maintain the pluralism of the present approach with a variety of funding agencies. The flexibility and local innovation of that approach are major strengths of the U.S. system of science and technology. At the same time, the agencies involved should be guided by a shared strategic view of future needs.

Recommendation 5. Stable funding should be provided for those undersea vehicle systems that are viewed as national assets.

APPENDICES

APPENDIX A

Biographical Sketches of Committee Members

J. Bradford Mooney, Jr., NAE (*chair*), is former president of Harbor Branch Oceanographic Institution and an ocean engineering and research management consultant to universities and industry. He retired from the U.S. Navy as rear admiral in 1987. His Navy career included assignments as chief of naval research, oceanographer of the Navy, and naval deputy to the National Oceanic and Atmospheric Administration (NOAA). His extensive prior experience in submarines and submersible vehicles included roles as a Navy deep submergence vehicle hydronaut; pilot of the *Trieste II* vehicle in the successful 1963 search for the sunken U.S. Navy submarine *Thresher* at a depth of 3,500 meters; coordinator of diving and submersible operations below 100 meters in the search and recovery of an H-bomb off Palomares, Spain, in 1966; officer in tactical command of a task force involved in the successful recovery of a classified object from a depth of 5,000 meters in the mid-Pacific in 1972; and founder of the Navy's first fleet operational deep submergence command. He received his B.S. degree from the U.S. Naval Academy and pursued postgraduate management studies at George Washington and Harvard Universities. Admiral Mooney was a member of the Marine Board and the Mine Countermeasures Study of the Naval Studies Board. He also has served on other National Research Council (NRC) boards.

John R. Apel recently retired from the Applied Physics Laboratory of Johns Hopkins University, where he was the principal staff physicist and chief scientist at the Milton S. Eisenhower Research Center. Previously, Dr. Apel served as director and supervisory oceanographer at NOAA's Pacific Marine Environmental Laboratory and earlier was director at NOAA's Atlantic Oceanographic and Meteorological Laboratories' Ocean Remote Sensing Laboratory. Also during his directorship with NOAA's Pacific Laboratory, he was affiliate professor of oceanography and affiliate professor of atmospheric sciences at the University of Washington, Seattle. Dr. Apel's research interests focus on geophysical fluid dynamics and nonlinear internal waves, Gulf Stream waves and instabilities, ocean remote sensing, and electromagnetic scatter from the ocean surface. He has served on many NRC committees and boards, including the NRC's Committee on Earth Sciences (June 1989 through June 1991), as well as on several satellite oceanography planning committees for NASA and the Office of Naval Research. Dr. Apel received his B.S. and M.S. degrees in physics from the University of Maryland and a Ph.D. in electrical engineering from Johns Hopkins University.

Robert H. Cannon, Jr., NAE, is Charles Lee Powell Professor (and was chairman 1979–1991), Department of Aeronautics and Astronautics, Stanford University. Previously, he was professor of engineering and chairman, Division of Engineering and Applied Science, California Institute of Technology, and associate professor at the Massachusetts Institute of Technology (MIT). Dr. Cannon also served as assistant secretary of transportation and chief scientist, U.S. Air Force. Earlier in his career, he contributed to advances in aircraft autopilots and contributed to gyro and stable-platform development for the polar voyages of the U.S. Navy submarines *Nautilus* and *Skate*. He received his B.S. degree from the University of Rochester in 1944 and D.Sc. from MIT in 1950, after serving as a naval officer. His research concerned a hydrofoil sailboat that held the unofficial world speed record for sailing. Dr. Cannon was a member of the NRC's Ocean Studies Board (until June 1995), has served on several other NRC boards, and was chairman of the Assembly of Engineering. He is currently working with Ph.D. students conducting experimental research studies in human strategic-level control of free-flying robots in space and under the ocean.

John R. Delaney is director of the Volcano Systems Center and professor of oceanography at the University of Washington, Seattle. He began his career in the mining industry searching for base and precious mineral deposits in Canada and the Western United States. He joined the faculty in Seattle in 1977 and has been chief scientist on more than 20 major seagoing expeditions, using a number of undersea vehicles to study deep sea volcanic activity at mid-ocean ridges. In 1987, Professor Delaney led an National Academy of Sciences workshop on interdisciplinary research related to spreading centers on the seafloor; the National Science Foundation-supported Ridge Initiative grew out of that event, and Professor Delaney served as the first chair of the steering committee from 1988 to 1992. His current research is focused on establishing long-term seafloor observations. He received a B.A. from Lehigh University, an M.S. from the University of Virginia, and a Ph.D. from the University of Arizona.

Norman B. Estabrook of Science Application International Corporation was formerly director of Ocean Engineering for the Naval Command, Control, and Ocean Surveillance Center, Research, Development, Test and Evaluation Division Naval Research and Development Section (NR&D), in San Diego, California. At NR&D, he was responsible for leading 88 engineers and scientists in the development of noncrewed vehicles systems, fiberoptic microcable systems, work systems, mine countermeasures vehicles, surveillance arrays, and related technologies. Mr. Estabrook has had more than 30 years of service involving developing, installing, and field testing many types of undersea systems, including the Navy's SEA LAB III, the acrylic-hulled manned submersible *Makakai*, research on the Navy's underwater work systems package, engineering support of Submarine Development Group One (the Navy's advanced manned submersible field test operation), and project management of the Advanced Unmanned Search System (autonomous vehicle). His experience largely parallels that of the late Howard Talkington, whom he replaced on this committee. Mr. Estabrook received an M.S. in aerospace engineering from the University of Southern California and a B.S. in engineering from the University of California at Los Angeles. He has completed predoctoral studies at the University of Hawaii. He is author of more than 15 major technical papers on ocean-engineering topics.

Larry L. Gentry recently retired from Lockheed Martin Marine Systems, where he was responsible for undersea vehicle programs and the development of advanced energy systems, materials, and sensor technologies for underseas equipment, including numerous advanced technologies used in developing underwater vehicles unoccupied by humans. Mr. Gentry has more than 30 years of experience in subsea and offshore engineering and program management, including applying aerospace technologies to the marine environment. His recent activities have focused on developing autonomous vehicles for the Navy and the Defense Advanced Research Projects Agency (DARPA). Earlier work involved developing seafloor work and subsea maintenance vehicles for use by the offshore oil industry, as well as developing seafloor cable laying systems and pipeline pull-in systems. He began his marine career as a commissioned officer and a diver in the U.S. Navy. Mr. Gentry received a B.S. from Oregon State University and an M.S. from San Jose State University, both in electrical engineering. He was a member of the Marine Board until June 30, 1992.

James R. McFarlane is president and founder of International Submarine Engineering (ISE). Since 1975 he has directed the design, construction, and operation of tethered and untethered remotely operated vehicles (ROVs) and the development of autonomous vehicles. Previously, he served as vice president of engineering and operations of International Hydrodynamics, where he was responsible for ROV development, construction, testing and operations. Prior to that, his 18-year career as an officer in the Canadian Armed Forces included assignments as senior structural engineer and staff officer on the staff of the Canadian Naval Submarine Technical Representatives, and as project manager for the SDL-1 diver lockout submersible. He is a founding member of the Canadian Academy of Engineering. He has authored numerous technical papers on submarines, human-occupied submersibles, ROVs, and autonomous vehicles. He received his B.Sc. degree in mechanical engineering from the University of New Brunswick, an M.S. in naval architecture and marine engineering, and a degree in naval engineering from Massachusetts Institute of Technology. He received honorary Doctor of Engineering degrees from Canada's Royal Military College and the University of Victoria, an honorary Doctor of Military Science from Royal Roads Military Academy, and an honorary Doctor of Science from the University of New Brunswick.

Andrew L. "Drew" Michel is the founder of ROV Technologies, Inc., which provides engineering services in the area of remote underwater intervention to oil companies and other major organizations worldwide. He has more than 30 years of experience with ROV systems and associated technology. As vice president and manager of a large undersea service company from 1966 to 1986, he was responsible for some of the earliest development and operations of electronics and ROV systems in commercial use. In his current position as technical director of ROV technologies, he supervises the work of engineers and ROV specialists on the nation's deepest oil and gas projects. He is chairman of the Marine Technology Society ROV Committee and a senior member of the Institute of Electrical and Electronic Engineers.

Bruce H. Robison is senior scientist and science chairman of the Monterey Bay Aquarium Research Institute. His research interests are focused on deep sea ecology and applying advanced submersible technology to oceanographic research. A qualified manned submersible pilot, he has dived in 12 different research submersibles and is a regular user of ROVs. He led the *Deep Rover* expedition, the first program to use submersibles to study California's

Monterey Submarine Canyon, in 1985. He received his B.S. degree from Purdue University; an M.S. from the College of William and Mary; and Ph.D. from Stanford University. He was also a postdoctoral fellow at Woods Hole Oceanographic Institution.

Mary I. Scranton is a professor at the State University of New York, Stony Brook, where she has been associated since 1979. Previously, she was a National Academy of Sciences/ National Research Council Resident Research Associate at the Naval Research Laboratory. Her research interests focus on the interaction between biological and chemical processes in the ocean, particularly on the initial and final stages of carbon degradation in water and sediments. She was a member of the *Alvin* Review Committee (renamed the Deep Submergence Science Committee), operating under the University-National Oceanographic Laboratory System. In 1994, she was the first woman to dive in the Navy DSV NR-1. She received her B.A. degree in chemistry from Mount Holyoke College and a Ph.D. in oceanography from the Massachusetts Institute of Technology/Woods Hole Oceanographic Institution Joint Program.

Peter H. Wiebe is senior scientist and department chairman at the Woods Hole Oceanographic Institution, where he has held various positions since 1969. In 1968 and 1969, he was a postdoctoral fellow at Hopkins Marine Station, Stanford University. His research interests include the small-scale spatial distribution of oceanic zooplankton and the transfer of energy to deep-sea populations by sinking large particles of particulate organic matter from surface waters. He received his B.S. degree in zoology/mathematics from Northern Arizona University, Flagstaff, and his Ph.D. in biological oceanography from the Scripps Institution of Oceanography.

Dana R. Yoerger is an associate scientist at the Deep Submergence Laboratory, Department of Applied Physics and Engineering, Woods Hole Oceanographic Institution. His major interests include underwater vehicles and manipulators. He is a principal in the design and application of *Argo/ Jason*, a telerobotic system designed for seafloor survey, and *ABE*, an autonomous vehicle used for long-term monitoring of the deep ocean. He has participated in numerous oceanographic cruises, including the discovery of the *Titanic*, the full-scale dynamic testing of the *Argo* system, and the deep-ocean deployment of the *Jason* vehicle and manipulator. He has published papers on vehicle and tether dynamics, the application of modern nonlinear and adaptive control techniques to underwater vehicle operation, supervisory control methodologies, and underwater manipulator design and performance. He received his B.S., M.S., and Ph.D. degrees in mechanical engineering from the Massachusetts Institute of Technology.

APPENDIX B

Foreign Developments

A variety of motivations drive international developments in undersea vehicles, and national government sponsorship of these developments varies from country to country. As in the United States, marine research, ocean mapping, minerals, communications, oil and gas, fisheries, and national security are the primary interests of international ocean technology. Japan, France, Canada, Russia, and the United Kingdom have the most aggressive ocean technology development programs outside the United States.

JAPAN

While Japan's ocean interests focus primarily on the "traditional" maritime sectors like fisheries and marine transportation, this nation also has the foremost undersea vehicle development program (Okamura, 1990). Because there is very little defense research and development (compared to the major military powers) driving technology in Japan, "headline projects" have instead provided national technology focus. The Japanese have adopted a systematic strategy to "conquer inner space" by progressively penetrating the ocean. *Kaiko* reached the deepest part of the ocean in March 1995. The Japan Marine Science and Technology Center (JAMSTEC), Ministry of International Trade and Industry, the Ministry of Transportation, the Ministry of Construction, the Advanced Robot Technology Research Association, Overseas Communication Japan, and the Japan Deep Sea Technology Association are among the many agencies and organizations involved in Japan's coordinated ocean technology effort. This effort includes deep submergence vehicles (DSVs), remotely operated vehicles (ROVs), and autonomous undersea vehicles (AUVs) (Asakawa et al., 1993). Japan has also aggressively pursued international agreements and has a number of ongoing collaborative projects with other countries (JAMSTEC, 1992).

Outside of the key "headline projects," Japan's interest is in technologies concerned with earthquake prediction and undersea resources, including submarine minerals and oils. The objective of these programs is more serious in regard to minerals, particularly in view of the United Nations (UN) Marine Law Treaty (in effect as of November 1994) that prescribes that mining concessions to exploit deep-seabed minerals should be provided only to companies with the necessary exploiting technologies. The UN will assess the engineering qualifications of the applicant. Because Japan has applied for mining concessions off the Hawaiian Islands, technologies advanced enough to qualify Japan as a developer of submarine minerals are needed.

Although the Japanese developed some occupied and unoccupied submersibles as early as the late 1950s, beginning in the early 1980s, they systematically developed increasingly deep-diving vehicles. The series began with a DSV, *Shinkai 2000* (2,000 meters), in 1981. The 3,000-meter ROV, *Dolphin 3K*, was put into service in 1987, followed by the DSV *Shinkai 6500* (6,500 meters), which became operational in 1989 and is currently the deepest-diving manned submersible in the world. The most recent addition to the Japanese vehicle assets is *Kaiko*, an ROV that cost approximately $60 million, which assists and accompanies *Shinkai 6500*. *Kaiko* made its initial deep test dive in March 1994, going into the Mariana Trench off Guam island, where it reached a depth that nearly matched the *Trieste* record. Umbilical problems forced termination of the dive just two meters from the bottom. *Kaiko* did reach the deepest abyss in the Challenger Deep of the Mariana Trench at 10,912 meters, on March 24, 1995, where it filmed small fish. *Kaiko* is equipped to do seismological and biological research in addition to supporting *Shinkai* missions.

JAMSTEC is a joint government-industry organization responsible for undersea vehicle procurements, maintenance, and operations. JAMSTEC has dock space for the three mother ships, *Natsushima* (for *Shinkai 2000*), *Kaiyo* (for *Dolphin 3K*), and *Yokosuka* (for *Shinkai 6500* and *Kaiko*), as well as complete on-shore maintenance facilities for all the vehicles. A simulator is provided for submersible pilot and crew training as well as mission simulation.

RUSSIA AND UKRAINE

Two of the republics of the former Soviet Union (FSU) have undersea vehicle assets and developmental programs. In the FSU the emphasis has been on DSVs and AUVs rather than ROVs. There are more than 30 DSVs in the FSU, which is more than in any other area of the world. Many of these DSVs are in Ukraine because Sevastopol in the Crimea (now part of Ukraine) was a major USSR support base for these vehicles; most are still in operational condition. There is now little new development and few operations because of present political and economic conditions.

In Russia, the two *MIR* submersibles (6,000-meter depth capability) built in Finland and operated by the Russian Academy of Sciences, Shirshov Institute of Oceanology, are considered state of the art. The Shirshov Institute has four other DSVs, with two more under construction. However, the *MIR*s are probably the best known DSVs in the FSU, where there is now a growing interest in developing tourist submersibles and AUVs for export.

The FSU built more than 20 ROV systems during the past 20 years for a variety of applications (Given, 1991), including some that were towed from ships and were capable of depths to 6,000 meters. As a result of technology export restrictions from the West during the Cold War, AUVs development in the FSU has been and will continue to be hampered by the lack of modern computing technology. Nonetheless, engineers in the FSU have been developing AUV technology since the early 1970s. Even without advanced electronics and computers, they have achieved a number of practical successes and have accumulated significant operational experience with autonomous vehicles. At least one project is under way to sell AUVs on the international market (Given, 1991). The Institute for Oceanological Problems in Vladivostok has used several AUVs in numerous deep ocean search and recovery operations in the Pacific and Atlantic oceans and in the Norwegian Sea (Mooney et al., 1996).

The FSU has also developed advanced capability in titanium fabrication and titanium welding, and the development of composite and ceramic structures is under way. Funding shortages mean that little new development in the FSU is currently under way; however, former Soviet research and development institutions and facilities are increasingly open to foreign visitors (OTA, 1993). In addition, they are actively marketing service operations using their submersible assets. The two *MIR*s have been used by companies and organizations in the West. Although this may or may not contribute directly to U.S. technological capabilities, the FSU will likely have technology to offer other countries, and the U.S. may adopt specific technologies of interest.

FRANCE

France, primarily the French Institute for Research and Exploitation of the Oceans (IFREMER), continues to have an impressive ocean technology program, which began in the 1950s. IFREMER is a public agency that has special status allowing it to function as a government-funded corporation while also conducting private industrial and commercial for-profit business. IFREMER's research assets include DSVs, ROVs, and AUVs. The 6,000-meter capable *Nautile* is probably the best known of the French DSVs; IFREMER is also developing a 6,000-meter ROV. Private companies in France, such as COMEX Industries, have developed a variety of commercial (i.e., nonmilitary) undersea vehicles. COMEX alone has designed and built more than 20 DSVs. Finally, the French navy has a long history of DSV development. Beginning in the 1950s, the bathyscaphes *FNRS-3* and *Archimede* were in service for nearly a quarter of a century before they were finally retired in the late 1970s. Today, the Navy operates the DSV *Griffon* in support of its deep-ocean missions.

UNITED KINGDOM

In the United Kingdom, the offshore oil and gas industry primarily drives ocean technology and focuses on ROVs and AUVs. In the past, private companies in the United Kingdom were the principal developers of undersea technologies, including vehicles and subsystems. They were also significant exporters of those technologies.

Within the government, the Defense Research Agency plays a key role in ocean technology work. The United Kingdom is not using or developing DSVs (although some were developed in the 1970s) with the single exception of a submarine rescue vehicle for the Royal Navy, which is an adaption of the *Slingsby LR5* built in 1978.

The United Kingdom is also involved in the European Community cooperative research program Marine, Science and Technology (MAST). MAST funds scientific and technological programs with parallel objectives. The overall goal of the programs is to contribute to establishing a *scientific and technical* [emphasis added] basis for the exploration, exploitation, management, and protection of the seas around Europe (MAST, 1990). Approximately 30 percent of MAST's $100 million budget for 1991 through 1994 was targeted for marine technology, including vehicles.

Greece, Portugal, Italy, and Denmark are also active in MAST. A large number of MAST's technology programs focus on sampling and measuring instrumentation, including optical plankton analysis systems, electrochemical instrumentation for the in situ determination of trace metals, the in situ acoustic characterization of suspended sediment, and anti-fouling coatings for submarine sensors.

The U.K. AUTOSUB AUV development program, which includes a proof-of-concept AUV and a 6,000-meter depth, long-duration AUV, parallels MAST. The AUV, called *Dolphin*, will be capable of transiting from the United Kingdom to the United States and collecting oceanographic data enroute (ITRI, 1994).

CANADA

For nearly 30 years, Canada has been a leader in the development and sale of undersea vehicles. One company (which ceased doing business in 1975) was the third largest builder (15 built) of DSVs in the world. Today, a company in Vancouver is the largest manufacturer of tourist submersibles, with more than 12 delivered throughout the world. Another Vancouver company proposed two tourist submarines in the early 1990s. In the ROV and AUV vehicle sectors, several Canadian companies have built a wide variety of ROVs, ranging from low-cost inspection vehicles to work ROVs. Several AUVs have also been built in Canada. In all, more than 200 undersea vehicles of various types have been produced in Canada (McFarlane, 1995). Canadian companies have also developed an aluminum-oxygen battery that has been tested in an AUV (Stannard et al., 1994).

Canadian scientists have also been evaluating the usefulness of ROVs as scientific platforms, and one DSV was used for university research. Canadian Armed Forces have used DSVs, ROVs, and AUVs for a variety of support missions. While this work represents only a few vehicles, it has been ongoing for more than two decades. The Canadian Hydrographic Service has also operated AUVs.

NORWAY

In Norway, the Defense Research Establishment developed a low-hydrodynamic-drag AUV powered by magnesium-seawater batteries (Apel, 1993). The original mission of the AUV was surveillance; however, it is now occasionally used for research. The AUV has operated under remote control via acoustic link out to ranges of 110 nautical miles with satisfactory results. Using low-voltage magnesium battery, a potential range of 1,100 to 1,200 nautical miles is possible. This battery has one of the highest specific energy specifications to date.

REFERENCES

Apel, J. 1993. Norwegian Defense Research Establishment. Mg-Seawater-Battery-Powered Autonomous Underwater Vehicle, January–September 1992. Circulated memo. Laurel, Maryland: The Johns Hopkins Applied Physics Laboratory.

Asakawa, K.J., J. Kohima, Y. Ito, Y. Shirasaki, and N. Kato. 1993. Development of autonomous underwater vehicle for inspection of underwater cables. Pp. 208–216 in Proceedings, Underwater Intervention '93 held January 18–21, 1993 in New Orleans, Louisiana. Washington, D.C.: The Marine Technology Society.

Given, D. 1991. Underwater technology in the USSR. Oceanus 34(1):67.

ITRI (International Technology Research Institute). 1994. World Technology Evaluation Center Program. Pp. 72–73 in World Technology Evaluation Center Panel Report on Research Submersibles and Undersea Technologies. Baltimore, Maryland: Loyola College of Maryland.

JAMSTEC (Japan Marine Science and Technology Center). 1992. Pp. 1–52 in Long-term Plan of Japan Marine Science and Technology Center. Yokosuka, Japan: JAMSTEC.

McFarlane, J. 1995. Personal communication to Donald W. Perkins, April 14, 1995.

MAST (Marine Science and Technology in the United Kingdom). 1990. P. 145 in Report of the Coordinating Committee for Marine Science and Technology (CCMST). London: Her Majesty's Stationery Office.

Mooney, J.B., H. Ali, R. Blidberg, M.J. DeHaemer, L.L. Gentry, J. Moniz, and D. Walsh. 1996. World Technology Evaluation Center Program. World Technology Evaluation Center Panel Report on Submersibles and Marine Technologies in Russia's Far East and Siberia, in press. Baltimore, Maryland: Loyola College of Maryland, International Technology Research Institute.

Okamura, K. 1990. Ocean technology in Japan: Recent advances, future needs and international collaboration. Journal of the Marine Technology Society 24(1):32–47.

OTA (Office of Technology Assessment). 1993. P. 15 in Statement of Peter A. Johnson before a Hearing of the Subcommittee on Mineral Resources and Development, Senate Committee on Energy and Natural Resources on November 4, 1993.

Stannard, J.H., G.D. Deuchars, J.R. Hill, and D. Stockburger. 1995. Sea trials of an aluminum/hydrogen peroxide unmanned underwater vehicle propulsion system. Pp. 181–191 in Proceedings Manual, AUVS '95 held July 10–12, 1995 in Washington, D.C. Arlington, Virginia: Association of Unmanned Vehicle Systems International.

APPENDIX C

Development of Deep Submersible Vehicles in the United States: 1958–1994

Date	Submersible	Status
1958	***TRIESTE***	Retired, 1964
1959	—	—
1960	—	—
1961	SPORTSMAN 300	Retired (ca 1965)
1962	PC3-X	Retired (ca 1978)
1963	PC-3B	Retired (ca 1967)
	SPORTSMAN 600	Retired (ca 1967)
	STAR I	Retired (ca 1966)
1964	ASHERAH	Retired (ca 1970)
	ALUMINAUT	Retired (ca 1974)
	ALVIN	Active (modified 1973)
	DEEP JEEP	Retired (ca 1967)
	PERRY PC 3A (2 bit)	Retired, 1975
	TRIESTE II	Retired, 1966
1965	**DEEP STAR 4000**	Retired (ca 1972)
1966	***HIKINO***	Retired (ca 1968)
	MORAY	Retired (ca 1969)
	STAR II	Retired, 1991
	STAR III	Retired (ca 1974)
	TRIESTE II (No. 2)	Retired, 1982
	KITTREDGE K-250	(41 built from 1966–1980; most are retired)
1967	DEEP QUEST	
	PAULO I (SEA OTTER)	Retired (ca 1988)
		Inactive
1968	BEAVER	Retired (ca 1985)
	BEN FRANKLIN	Retired (ca 1971)
	DEEP DIVER (PLC4)	Retired (ca 1972)
	DOWB	Retired (ca 1972)
	SHELF DIVER	Inactive
	TURTLE	Active (modified 1985)
	NEKTON ALPHA	Inactive
1969	DEEPSTAR 2000	Retired (ca 1977)
	KUMUKAHI	Retired (ca 1973)
	NR-1	Active
	SNOOPER	Active
1970	***DSRV-1***	Active
	NEKTON BETA	Inactive
	NEMO	Retired
	SEA CLIFF	Active (modified 1982)
	SURVEY SUB 1 (PC-9)	Retired (ca 1974)

Date	Submersible	Status
1971	***DEEP VIEW***	Retired (ca 1973)
	DSRV-2	Active
	MAKAKAI	Retired (ca 1975)
	NEKTON GAMMA	Active
	SEA-LINK I	Active
1972	OPSUB (PERRY)	Retired (ca 1974)
	SEA RANGER	Retired (ca 1980)
1973	—	—
1974	PC-14C-1 CLELIA (ex DIAPHUS)	Active
1975	SEA-LINK II	Active
	PERRY PC-14C-2	Active
1976	PISCES VI	Inactive
1977	—	—
1978	PIONEER I	Inactive
1979	—	—
1980	—	—
1981	—	—
1982	DELTA	Active
1983	PERRY PC-1805	Inactive
1984	DEEP ROVER	Active

NOTES: This table does not include ROVs or AUVs. DSVs in bold italics were developed by U.S. Navy. This table does not include one-off and *backyard* submersibles. The line between *inactive* and *retired* is not precise. In general, *inactive* refers to DSVs that can be put back into service without excessive restoration. *Retired* designates submersibles that cannot be put back into service, have been given to museums, junked, or are similarly unavailable. During years with dashes, no DSVs were put into service. The submersibles listed were built in the United States and worked at least part of their lives for U.S. companies. Not included are DSVs built by U.S. companies (primarily Perry) and exported or foreign-built DSVs now used by U.S. companies.

APPENDIX D

U.S. Government Agencies That Own and/or Use Submersibles[a]

Agency	Submersible	Owned	Leased	Purposes
U.S. Navy	DSVs	X	X	General mission support
	ROVs	X	X	General mission support
	AUVs	X		Experimental development
U.S. Army Corps of Engineers	ROVs	X	X	Inspection of dams and other underwater structures
U.S. Army Ballistic Defense Command	DSVs		X	Recovery of missile parts in splash-down area of range
U.S. Air Force	DSVs		X	Recovery of missile parts in splash-down zone of missile range
Advanced Research Projects Agency	AUVs	X	X	Experimental development
National Aeronautics and Space Administration	ROVs	X		Development of telepresence vehicle controls for missions to other planets
Environmental Protection Agency	DSVs		X	Inspection of seafloor dump sites
National Oceanic and Atmospheric Administration[b]	DSVs		X	General oceanographic research
	ROVs	X	X	General oceanographic and fisheries research
U.S. Department of the Interior[c]	DSVs		X	Research in Crater Lake, Oregon, for National Park Service
	ROVs		X	
U.S. Drug Enforcement Administration	ROVs	X		Drug interdiction
U.S. Coast Guard	ROVs	X		Maritime law enforcement
U.S. Customs Service	ROVs		X	Underwater inspection of ships and structures for contraband

[a]This listing illustrates the scope and diversity of U.S. federal agencies that have employed submersibles to support their missions. No attempt has been made to describe each vehicle used in the past 35 years, and there are probably more than are shown here. The intent is to show that a variety of government agencies have used submersibles.

[b]National Oceanic and Atmospheric Administration agencies that have used submersibles are: the National Undersea Research Program and the National Marine Fisheries Service.

[c]U.S. Department of the Interior agencies that have used submersibles are the U.S. Geological Survey, the Minerals Management Service, and the National Park Service.

APPENDIX E

Deep Submersible Vehicles in Service or Available Worldwide

Country	Submersible	Operator	Depth	Date Built or Modified
USA[a]	Alvin	Woods Hole Oceanographic Institution	4,500 m	1964/mod 1973
	Clelia	Harbor Branch Institution of Oceanography	350 m	1974
	Delta	Delta Oceanographics, Inc.	300 m	1982
	Diving Saucer	Jacques Cousteau Society	350 m	1959
	Nekton Gamma	Carribean Research Lab	300 m	1970
	NR-1	U.S. Navy	700 m	1969
	Pisces V	University of Hawaii	1,500 m	1973
	Sea Cliff	U.S. Navy	6,000 m	1964/mod 1982
	Sea-Link 1	Harbor Branch	800 m	1971
	Sea-Link 2	Harbor Branch	800 m	1975
	Snooper	Undersea Graphics	300 m	1969
	Turtle	U.S. Navy	3,000 m	1968/mod 1985
Russia[b]	Argos	P.P. Shirshov Institute	600 m	1975
	MIR I	P.P. Shirshov Institute	6,000 m	1987
	MIR II	P.P. Shirshov Institute	6,000 m	1987
	Osmotr	P.P. Shirshov Institute	300 m	1985
	Pisces XI	P.P. Shirshov Institute	2,000 m	1975
	Pisces VII	P.P. Shirshov Institute	2,000 m	1975
Ukraine[c]	Sever 2	Academy of Science	2,000 m	1969
	Sever 2 Bis	Academy of Science	2,000 m	1970
	Benthos I	Academy of Science	300 m	1969
	Benthos II	Academy of Science	300 m	1972
	Tinro-2	Academy of Science	400 m	1973
	Tinro-2 Bis	Academy of Science	400 m	1974
France[d]	Cyana	IFREMER	3,000 m	1970
	Deep Rover 2	Ellipse Program	1,000 m	1994
	Deep Rover 2	Ellipse Program	1,000 m	1994
	Nautile	IFREMER	6,000 m	1985
	Saga	IFREMER	600 m	1987
	SO-450-Vaimana	Papete, Tahiti Industrie et Tourisme Sous-Marin	450 m	1982
Japan	Shinkai 2000	JAMSTEC	2,000 m	1981
	Shinkai 6500	JAMSTEC	6,500 m	1987
Canada	Deep Rover	Can Dive	1,000 m	1984
	Pisces IV	Institute of Ocean Science	2,000 m	1972

Country	Submersible	Operator	Depth	Date Built or Modified
Romania[e]	SC-200	Institute of Oceanology	200 m	1979
	SM-358	Institute of Oceanology	300 m	1979
Switzerland	A.F. Forel	Piccard	500 m	1979
Germany	JAGO	Max Planck Institute	500 m	1990
Finland	SM 80/2	Finnish National Board of Waters and Environment	500 m	1990
South Korea[f]	Hae Yang 250	Korea Ocean Research and Development Institute	100 m	1988
Bulgaria[g]	PC-8	Institute of Oceanology	200 m	1981

SOURCE: Don Walsh, International Maritime, Inc.

[a]U.S. Navy submersibles primarily support Navy missions; however, significant amounts of time are provided to the U.S. marine science community each year.

[b]Because of uncertain economic conditions in Russia, few operations are being conducted at present. Several of these submersibles have been offered for sale.

[c]Because of uncertain economic conditions in Ukraine, few operations are being conducted at present. Several of these submersibles have been offered for sale.

[d]Saga is currently in storage. The SO-450 will be used in French Polynesia for scientific research and commercial operations.

[e]The operational status of these submersibles is unknown because of uncertain economic conditions in Romania.

[f]This Kordi submersible was built in Korea and is currently inactive.

[g]Because of uncertain economic conditions in Bulgaria, this submersible is believed to be inactive.

Acronyms

ADCP	acoustic Doppler current profiler	**LCROV**	low-cost ROV
ARPA	Advanced Research Programs Agency (an agency within the U.S. Department of Defense; see also DARPA)	**LLS**	laser line scanner
		NASA	National Aeronautics and Space Administration
ASW	antisubmarine warfare		
AUSS	Advanced Unmanned Search System	**NOAA**	National Oceanic and Atmospheric Administration
AUV	autonomous underwater vehicle or autonomous undersea vehicle	**NSF**	National Science Foundation
		NURP	National Undersea Research Program (an office within NOAA)
CCD	charge coupled devices		
CTD	conductivity, temperature, depth instrument	**ONR**	Office of Naval Research
		OTEC	ocean thermal energy conversion
DARPA	Defense Agency Research Programs Agency	**ROV**	remotely operated vehicle
DESSC	Deep Submergence Science Committee (a committee within UNOLS; see also UNOLS)	**SE&I**	systems engineering and integration
		UNOLS	University-National Oceanographic Laboratory System
DSRV	deep submergence rescue vehicle		
DSV	deep submersible vehicle	**USCG**	U.S. Coast Guard
EEZ	Exclusive Economic Zone	**UUV**	unmanned undersea vehicle
JAMSTEC	Japan Marine Science and Technology Center		
JOI	Joint Oceanographic Institutions, Inc.		